T0186614

Designing Engineering Structures Using Stochastic Optimization Methods

Editors

Levent Aydin
Department of Mechanical Engineering
Izmir Katip Celebi University, Izmir, Turkey

H Seçil Artem
Mechanical Engineering Department
Izmir Institute of Technology, Izmir, Turkey

Selda Oterkus
Department of Naval Architecture, Ocean and Marine Engineering
University of Strathclyde, Glasgow, UK

CRC Press
Taylor & Francis Group
Boca Raton   London   New York

CRC Press is an imprint of the
Taylor & Francis Group, an **informa** business

A SCIENCE PUBLISHERS BOOK

CRC Press
Taylor & Francis Group
6000 Broken Sound Parkway NW, Suite 300
Boca Raton, FL 33487-2742

First issued in paperback 2022

© 2020 by Taylor & Francis Group, LLC
CRC Press is an imprint of Taylor & Francis Group, an Informa business

No claim to original U.S. Government works

Version Date: 20200225

ISBN-13: 978-0-429-28957-6 (ebk)
ISBN-13: 978-0-367-51002-2 (pbk)
ISBN-13: 978-0-367-25519-0 (hbk)

DOI: 10.1201/9780429289576

Visit the Taylor & Francis Web site at
http://www.taylorandfrancis.com

and the CRC Press Web site at
http://www.crcpress.com

Preface

Among all aspects of engineering, design is the most critical in developing a new product. In real life situations, it is not always practical to use traditional methods. A systematic approach to design problems can accomplish much, and can be used by applying mathematical optimization processes. Stochastic optimization methods such as differential evolution, simulated annealing, and genetic algorithms are some of the effective ways in arriving at optimal designs.

There are very few books on optimization that include engineering applications because of limited study on it. Further, engineering applications are generally limited to well-known design problems. Most books comprise a collection of topics of advanced nature, therefore typical problems are not always apparent and remain theoretical. To address this, each chapter in this book is contributed by at least one academic and one industrial expert.

The present book is intended for researchers, institutes involved in design-related projects, graduate students of engineering, and industrial product designers who want to have a better understanding of the challenges of the design process. Several chapters can also be used for senior undergraduate courses in engineering-solid and fluid mechanics. It is also intended for practicing engineers who would like to learn more about the theory of design optimization.

A significant portion of the book emphasizes the neuro-regression approach to model the output parameters of the engineering systems in terms of inputs. This modeling approach combines the advantages of regression analysis and artificial neural network (ANN), and provides insights into the prediction capability of candidate models. The first three chapters are devoted to the background knowledge, including modeling, design of experiments, and modern optimization algorithms. They are designed such that readers do not misinterpret the theories and results, and in this way, it is possible to evaluate outcomes based on the concepts presented.

The book includes a useful review of mechanical engineering design optimization by stochastic methods. Many engineering cases have been discussed to facilitate a scientific basis for industrial design problems. It is also aimed at students and design engineers on complicated mathematical optimization procedures. Beyond a work of collected papers on a particular topic, this book clearly and explicitly describes the steps of a wide range of selected engineering design problems. It describes how engineering structures are systematically designed. It has many new engineering design applications based on stochastic optimization techniques. The design problems are selected from *automotive* (Chapters 4 and 5), *energy* (Chapters 6–10), *military, naval industries* (Chapters 10 and 11), *manufacturing process* (Chapters 12 and 13), and the topics of *fluids-heat transfer* (Chapters 14–16). For each design optimization problem described, sufficient background is provided to explain the steps. Implementations of *Mathematica* (a commercial software) are also given in the solution procedures of the prescribed optimization problems.

Levent Aydin
H Secil Artem
Selda Oterkus

Contents

CHAPTER 1
Mathematical Background

Levent Aydin,[1,*] *H Seçil Artem*[2] and *Selda Oterkus*[3]

Introduction

This opening chapter surveys the fundamental mathematical instruments that will appear throughout this book. Many researchers are going to be familiar with several of the topics presented here. The reader can evaluate which areas need to be studied further. The main areas covered in this book are the design of experiments, regression analysis, neuro-regression approach, and boundedness concept. The main goal of the chapter is to present the basic and modern approaches to the modeling and design of engineering structures in a concise manner. Figure 1.1 shows all these steps without breaking the order of the optimal design process for a system.

Design of Experiments (DOE)

Design of Experiments (DOE) is a helpful tool for finding new processes, trying to learn more about the existing processes, and optimizing them for an excellent performance. In this section, we present and discuss some DOE techniques. As mentioned before, the list of techniques considered is far from complete as this section's aim is to introduce the reader to the topic by showing the main techniques used in practice. It is crucial to choose adequate statistical tools to analyze the data available, since the

[1] İzmir Katip Çelebi University, Department of Mechanical Engineering, İzmir, Turkey.
[2] Izmir Institute of Technology, Department of Mechanical Engineering, İzmir, Turkey.
    Email: secilartem@iyte.edu.tr
[3] University of Strathclyde, PeriDynamics Research Centre, Department of Naval Architecture, Ocean & Marine Engineering, Glasgow, Scotland.
    Email: selda.oterkus@strath.ac.uk
* Corresponding author: leventaydinn@gmail.com

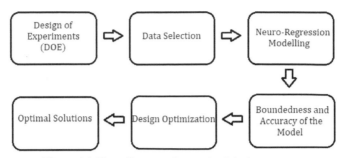

Figure 1.1 Flow diagram of an optimal design process.

results can be significantly affected by noise. Replication, randomization, and blocking are the fundamental principles of statistical methods in DOE. Replication is the experiment's repetition to achieve a more accurate result and to reduce the experimental error. Randomization identifies the random order in which the experiment runs are to be carried out. Blocking is intended to isolate a known systematic bias effect and inhibit the main effects from being obscured [1, 2]. Due to the number of crunchers concerned and the use of complicated statistical jargon engineers generally do not like to employ DOE [2]. In manufacturing processes, experiments are being carried out to enhance our knowledge and understanding of them. The relationships between the core factors of the inputs and the behaviors of the output can, therefore, be examined [2, 3]. One of the comprehensive strategies in engineering companies promoted by plenty of engineers is One-Variable-At-a-Time (OVAT). In this approach, it is changed one parameter at a time, with all other factors fixed throughout the experiment. The results, however, are unreliable, wasteful, and may offer the processes a misleading conclusion. If a particular attribute of a component is affected by several factors, then the best choice is DOE [2, 3, 4]. The engineer frequently makes systematic changes in the input parameters and specifies how well the output performance varies. It is known that all parameters have not the same impact on the results. The aim of a deliberate design is, therefore, to realize which process parameters tend to affect the output more and then to identify the best levels for the factors [1, 2, 5]. This approach provides high process efficiency, more stable results, low manufacturing costs, and, saves time for the researchers.

DOE Techniques

The selection of a DOE approach relies on the experiment's goals and the number of factors to be considered. In this section, the most widely used approaches have been listed and explained briefly. These approaches are Randomized Complete Block Design, Full Factorial, Fractional Factorial,

Central Composite, Box-Behnken, Taguchi, Latin Hypercube, and D-Optimal Design [1].

Randomized Complete Block Design

The distribution of treatment for experimental components is not strictly limited. In practice, however, there are situations in which the experimental data varies relatively widely. In such situations, the design made in relation is called a Randomized Complete Block Design (RCBD). The main goal of blocking is to minimize the variability between experimental units within a block and to maximize the variation between blocks.

Advantages of the RCBD

1. It is possible to remove the treatments or replicates from the analysis.
2. Several treatments can be more often replicated than the others.
3. There is no strict limitation on the number of treatments or replicates.
4. Even if the experimental error is not homogeneous, there can still be valid comparisons [2, 6].

Disadvantages of the RCBD

1. There exists a smaller error on *df* for a small number of treatments.
2. If the number of treatments is enormous and there is a considerable variation between experimental units, it is possible to obtain a significant error term.
3. RCBD is not very good on the efficiency of the experiment when there is missing data.

Full Factorial Design

It is generally known that the DOE method most frequently used in manufacturing industries are full and fractional factorial designs at two and three levels. Factorial designs might allow a researcher to explore a response consistent on the impact of the variables. A factorial design can be separated as full or fractional factorials. An experimental design that every factor setting occurs with every other one is a full factorial design. If the amount of factors is five or higher, a full factorial design needs a significant amount of runs and is not very useful. In these cases, a fractional factorial design is a better choice [2, 3].

Fractional Factorial Design

Researchers usually do not have sufficient time, money, and funding to conduct full factorial experiments. If some higher-order relationships are not essential, the primary effects and two-order interactions can be acquired by operating only a fraction of the full factorial experiment. A fractional factorial design is defined as a form of orthogonal array layout that enables researchers to explore significant impacts and the required impacts of relationships with a minimum amount of exercises or experimental runs [2, 3, 6, 7].

Central Composite Design

The most popular response surface design is a central composite design, which creates a factorial design. Five factorial levels are desired for a central composite design. One of the most critical advantages is that the corner points are checked, if it is shown that curvature is not substantial, then it is accomplished. If the curvature exists, the primary task is to generate the star runs [4].

Box-Behnken Design

Box-Behnken Design is based on the cube edge midpoints rather than the corner points, lead to fewer runs; however, apart from the Central Composite Design, all runs must also be done. It should be noted that only three-factor levels are appropriate in The Box-Behnken Design. It has advantages if the curvature stated in the screening experiment is likely necessary [2].

Taguchi Design

Taguchi methods, or sometimes called robust design methods, are statistical methods. The primary purpose is to keep the output fluctuation minimal even in the appearance of noise. The technique significantly increases the efficiency of engineering. The Taguchi design helps to guarantee product quality by deliberately considering the noise factors and the amount of mistake in the area. This approach centers on enhancing the primary function of the design process, thereby promoting flexible designs [2, 6].

Advantages

1. It is simple and easy to use in several other engineering circumstances, enabling it as a robust yet straightforward tool.
2. It underlines, within some qualification constraints, a mean production characteristic value comparable to the final value instead of just a value, thus enhancing the quality of the product.
3. Without an impractically large number of testing, it enables the investigation of many distinct variables.

Disadvantages

1. Precisely that the acquired findings are only comparative and do not specify which variable has the most significant impact on the characteristic value of the product.
2. It can not be used in all interactions among all the parameters, since orthogonal arrays do not examine all parameter combinations.
3. It is hard to account for parameter interactions.
4. It is offline and hence improper for a procedure that changes dynamically, as in a computer simulation.

Latin Hypercube Design

It is a method for generating a near-random sample of parameter values from a Multidimensional Distribution and is also the generalization of Latin Square Concept to an arbitrary number of dimensions. In this approach the first step is, to identify how many sample points to get an address and through which row and column the sample point was drawn for each sample point. Latin Hypercube method guarantees that the set of random numbers represents the real fluctuation, while standard random sampling is only a set of random numbers with no ensures [1–3].

Optimal Design (D-Optimal)

It is a computer-aided design that includes the finest portion amongst all feasible experiments. Software tools may also have distinct processes to generate D-optimal designs because the final design may vary on the tool to be used [2, 7]. A selection method creates the finest design based on a chosen factor and a specified amount of test runs. This approach is especially helpful if classical design methods are not being used. These situations are:

1. If supplies or factor configurations are restricted.
2. If it is necessary to reduce the number of design runs.
3. When using the operation and mixing variables in the same design.
4. When the experiments already carried out must be included.
5. If the region of the experiments is unstable [2, 4, 5].

Mathematical Modeling

When examining the literature, studies performed for the intent of engineering optimization, some inadequate approaches were recognized and listed below:

i) Since, it is necessary for consideration the interaction between all experimental and constructional parameters, from the viewpoint of optimization, updating in the one input with preserving the other constant is not a satisfying description, and this approach leads to disregarding the nonlinear influence of input variables.

ii) Most of the studies on modeling and optimization select only one or two traditional regression models as an objective function for the optimization problem. Further calculation of the R^2 value of the model for experimental studies is the main issue. However, a high value of R^2 itself does not describe the whole physical phenomena of the engineering process. The R^2 value is the proximity of the fitted model results to the experimental data. In other words, R^2 value does not always mean a good fit even if it is very high for the real systems. Besides, the model describes only the experimental data but not the fundamental behavior of the phenomena. Therefore, it is necessary to attempt new engineering modeling studies, including different regression forms and approaches.

iii) Besides these, another important feature of the engineering model function is that it should be bounded. Boundedness is relevant to realistic modeling of engineering systems, and it is known that all the engineering parameters are finite. Therefore, before the optimization step, it should be checked which of the selected models are also bounded under the engineering parameter intervals.

iv) Some of the available studies on the optimization of engineering systems do not take into account the reliability, sensitivity, and robustness of the algorithms. However, this is critical to reveal inherent behaviors of the stochastic search processes.

Because of these reasons, we introduce a novel approach to the modeling-design-optimization process in order to optimize the engineering input parameters through the book. Firstly, a detailed study on multiple nonlinear neuro-regression analysis, including linear, quadratic, trigonometric, logarithmic, and their rational forms for the prescribed problem (output) are performed. Secondly, boundednesses of the candidate models are checked to provide generating realistic values. Finally, the different direct search methods, including stochastic ones, are performed.

Mathematical modeling is the first step of the optimization process of engineering design problems. Therefore, the researchers try to distinguish mathematical models by utilizing Regression Analysis (RA), Response Surface Methodology (RSM), Finite Difference Technique (FDT), and Artificial Neural Networks (ANN). The obtained mathematical model also corresponds to the objective function of the prescribed optimization problem.

Neuro-regression Approach

In the modeling stage, a hybrid technique that combines the strengths of regression analysis and Artificial Neural Network (ANN) is used to improve the accuracy of the predictions. In this approach, all data is divided into two sets such that 80% and 20% of the given data and the first portion of the data is used for training; the second portion is for testing. In the training process, the aim is to minimize the error between the experimental and predicted values by adjusting the regression models and their coefficients. The test step is then conducted to achieve the outcomes of the prediction by minimizing the influences of regression model discrepancies, and this helps to bring insight into the candidate model's predictive capacity. Second, it is essential to verify candidate model's boundaries for prescribed values to demonstrate if the model is realistic or not. In this regard, the maximum and minimum values of the models in the given interval for each design variables are calculated after obtaining the appropriate models in terms of R^2 training and testing. As a result, chosen models meet the numerous criteria required for reality.

Nonlinear Regression Analysis

Nonlinear regression models are those that are not linear in their parameters and can be used for three different purposes [8]:

1. To test (or compare the hypothesis) the validity of the model,
2. Characterize the model (i.e., to estimate the parameters),
3. Predict the behavior of the system (interpolation and calibration).

The nonlinear regression model can be written as specified in the following equation:

$$Y = f(X, \beta) + \varepsilon \qquad (1.1)$$

Where:
X is a vector of p predictors, β is a vector of k parameters, f(–) = a known regression function,

- ε = an error term.

For nonlinear regression, mathematical modeling processes can be carried out systematically considering the essential features mentioned in the following items:

a) Nonlinear regression is more flexible than linear regression because the function does not need to be linear or linearizable. Therefore, the nonlinear regression phenomenon provides a wide selection of options to match the data.

b) Nonlinear regression may be more appropriate than the use of transformations and linear regression where the f function can be linearized.

c) Nonlinear regression requires a knowledge of the function f such as polynomial, trigonometric, exponential, which requires a thorough understanding of the process being studied. Linear regression models, on the other hand, are suitable for process estimations that are roughly certain of the relationship between input and output but do not require precise clarity.

d) Since nonlinear regression models contain the most general mathematical expressions, it is not possible to write functionally generalized states. However, a few basic types of models used in the engineering fields can be expressed as follows:

Examples of nonlinear equations are:

$$y = a_0 + a_1 x + a_2 x^2 + \dots a_n x^n \tag{1.2}$$

$$y = a_0 + a_1 e^x + a_2 e^{x^2} + \dots a_n e^{x^n} \tag{1.3}$$

$$y = a_0 + a_1 \sin(x) + a_2 \sin(x^2) + \dots a_n \sin(x^n) \tag{1.4}$$

$$y = \frac{a_0 + a_1 x + a_2 x^2 + \dots a_n x^n}{b_0 + b_1 x + b_2 x^2 + \dots b_n x^n} \tag{1.5}$$

At this stage, the multivariable states of the above-mentioned model types containing more than one input can also be derived with similar logic. Another critical point is that, for example, special functions such as Bessel, Laguerre, Lambert, and Gamma or different combinations of elementary functions can be selected as model structures with a broader understanding of the families of mathematical functions.

References

[1] Cavazzuti, M. 2013. Optimization Methods From Theory to Design: Scientific and Technological Aspects in Mechanics. Springer, Berlin.

[2] Antony, J. 2014. Design of Experiments for Engineers and Scientists. 2nd Edition. Elsevier, London.

[3] Vecchio, R.J. 1997. Understanding Design of Experiments. Gardner Publications, USA.

[4] Montgomery, D.C., Runger, G.C. and Hubele, N.F. 2010. Engineering Statistics. 5th Edition. John Wiley and Sons, USA.

[5] Antony, J. 1998. Some key things industrial engineers should know about experimental design. Logistics Information Management 11(6): 386–392.

[6] Gunst, R.F. and Mason, R.L. 1991. How to Construct Fractional Factorial Experiments. ASQC Quality Press, Milwaukee, Wisconsin, USA.

[7] Montgomery, D.C. 2001. Design and Analysis of Experiments. John Wiley and Sons, USA.

[8] Turhan, F. 2017. 1100 serisi alüminyum malzemelerde tig kaynağı ile oluşan kaynak dikiş geometrisinin optimizasyonu. M.S. Thesis, İzmir Katip Çelebi University, İzmir.

CHAPTER 2

Stochastic Optimization Methods

H Irem Erten,[1] *H Arda Deveci*[2] *and H Seçil Artem*[1,*]

Introduction

Finding an approximation of an optimal solution for a function which is defined on a subset of finite-dimensional space is one of the most common problems in applied mathematics. In combinatorial optimization problems which are crucial for most machine learning approaches, there are some objective functions that are supposed to be optimized to find an approximation of an optimal solution. Fifty years ago, there were a lot of numerical optimization procedures for these optimization problems; most of them were deterministic (traditional optimization techniques). However, with the development of computer technology, stochastic methods (non-traditional optimization techniques) have become the essential tools for areas such as engineering, science, business, and statistics. These methods are relatively the latest available and popular because of the particular properties which the deterministic algorithm does not have [1, 2]. For instance, stochastic methods always include probability, such as according to the random distribution of rainfall and water usage, in a reservoir, predicting the water level periodically or forecasting the number of dropped connections for a communications network based on randomly variable but appropriate constant bandwidth. On the contrary, deterministic methods include probability under no circumstances; and outcomes take place based on exact input values [3].

[1] Izmir Institute of Technology, Department of Mechanical Engineering, İzmir, Turkey.
  Email: erteniremm@gmail.com
[2] Erzincan Binali Yıldırım University, Faculty of Engineering, Department of Mechanical
  Engineering, Erzincan, Turkey.
  Email: hadeveci@erzincan.edu.tr
* Corresponding author: secilartem@iyte.edu.tr

Stochastic optimization is the process based on minimizing or maximizing the value of a statistical or mathematical function when one or more than one input parameters depends on random variables. The randomness may be present as either noise in measurements or Monte Carlo randomness in the search procedure, or both [1, 2].

Many industrial, economic, biological, and engineering problems can be accepted as stochastic systems, such as communication area, signal processing, geography, aerospace, banking. In these systems, stochastic optimization is appropriate in order to solve decision-making problems, and many researchers have considered stochastic optimization methods in solving these problems. For instance, Yan et al. [4] suggested a qualitative and quantitative combined modeling specification depending on a hierarchical model structure framework which consists of the meta-meta model, the meta-model, and the high-level model. The results of the study showed that the proposed method could comprehensively describe the complex system. Li and Zhang [5] studied the problem of uncertain stochastic linear-quadratic optimal control under the inequality constraints for the final state. In this study, they proved the Karush-Kuhn-Tucker (KTT) theorem with hybrid constraints, and then they obtained new types of Riccati equations. This equation provides the necessary conditions for an optimal linear state feedback control existence which is produced by KKT theorem. The design of a dynamic programming algorithm was achieved in order to solve the uncertain constrained stochastic linear quadratic subject. Aydın et al. [6] studied on the design of the dimensionally stable laminated composites by the use of the efficient global optimization method (EGO). The optimization problem of a composite plate under high stiffness and low coefficients of thermal and moisture expansion was solved. The proposed optimization algorithm in this study is experimentally verified. After the design and optimization processes were completed, failure analysis of the optimized composites has been performed using Tsai–Hill, Hoffman, Tsai–Wu and Hashin–Rotem criteria. Generic steps of stochastic optimizations for renewable energy applications were extensively examined by Zakaria et al. [7]. Furthermore, the positive and negative sides of the stochastic optimization were emphasized. Significant optimization methods belong to the stochastic optimization stages are emphasized.

In their study, Niamsup and Rajchakit [8] introduced the latest improvements and significant stochastic optimization methods. It is claimed that the stochastic optimization methods are always more efficient than the deterministic optimization techniques for social, economic, technical aspects of renewable energy systems. Niamsup and Rajchakit examined polytopic discrete-time stochastic functions in the interval time-varying delays using the parameter-dependent Lyapunov-Krasovskii functional combined with linear matrix inequality techniques and new criteria for the robust stability of the stochastic system were proposed.

Maggioni et al. [9] studied the problem which was encountered by a bike-sharing service provider who needed to manage a fleet of bikes over a set of bike-stations, each with given capacity and time-varying stochastic demand. We consider one-track bike-sharing systems with transshipment, multi-stage, and two-stage stochastic optimization models are proposed to specify the optimal number of bikes to appoint to every station at the beginning of the service. Finally, managerial insights are provided comparing the solution supplied in the real system with the solutions obtained by using the two-stage and the multi-stage models.

Gutierrez et al. [10] studied on how to cope with the problem of indefinite case in the optimal management of the hydrogen network of a petroleum refinery. A two-stage stochastic optimization method was used to analyze the effect of raw changes in the operation of the network, and they were analyzed on how to solve the hydrogen network problem to obtain feasible solutions with stochastic and deterministic solutions by using real plant data.

Khayyam et al. [11] proposed a stochastic optimization model for carbon fiber production in the carbonization process in order to reduce energy consumption in a proper range of fundamental mechanical properties. They developed processing operations, and fifty samples of fiber were analyzed for each set of operations, their tensile strength, and modulus. During the production of the samples, the energy consumption was monitored on the processing equipment, and the five distribution functions were examined in order to determine distribution functions which could best describe the mechanical properties distribution of filaments. The Kolmogorov–Smirnov test was performed in order to confirm the distribution goodness of fit and correlation statistics. The result of the study showed that the production quality could be predicted using the stochastic optimization models in the given range, and this method minimized the energy consumption of its industrial process.

Tifkitsis et al. [12] developed a stochastic multi-objective cure optimization methodology and performed it on the thick epoxy/carbon fiber laminates. Kriging method, which substitutes into Finite Element (FE) simulation, was used to construct a surrogate model for computational efficiency. Surrogate model and Monte Carlo were coupled and integrated into a stochastic multi-objective optimization framework depend on Genetic Algorithms. The results indicated that there exists a significant reduction of 40% in the temperature overshoot and cure time in comparison to standard cure profiles.

Genetic Algorithm (GA), Simulated Annealing (SA), Differential Evolution (DE), Particle Swarm Optimization (PSO), Ant Colony Optimization (ACO), Artificial Bee Colony (ABC), Markov Chain Monte

Carlo (MCMC), Tabu Search (TS), Harmony Search (HS), Grenade Explosion Method (GEM), Covariance Matrix Adaption (CMA) are the examples of stochastic optimization methods [1–4]. Researchers continue to improve, and to add new stochastic methods or both to the literature. In the following subsections, some commonly used stochastic optimization methods are briefly overviewed.

Genetic Algorithm

Genetic Algorithm (GA) is an adaptive search algorithm that mainly depends on evolutionary algorithm and generates high-quality solutions for some complex engineering and optimization problems. GA method is based on the basic unit of genetics and natural selection, which means species which can adapt to changes in their environment and can survive, reproduce, and hand down knowledge to the next generations.

GA codes the input parameters or design variables of the optimization problem in the solution typesettings of a finite length. It usually works with coding; in contrast, traditional optimization methods generally work with design variables or input parameters. Once the initial generation is created, this algorithm improves the generation using the most prevalent genetic operators that are (i) selection, (ii) crossover, (iii) mutation. The main steps of the algorithm are given in Figure 2.1 [2, 13, 14].

(i) **Selection Operator**

   The selection operator aims to refer to the individuals who have better fitness scores which enable them to pass their genes on to the consecutive generations [2, 13, 14].

(ii) **Crossover Operator**

   This operator is one of the most significant genetic algorithm operators which has an essential task of making new children in a reproduction operation. The crossover operator controls mating between individuals. After two individuals are selected by using selection operator, crossover areas are chosen at random, and then the genes which are at crossover sites are exchanged. In this way, a new individual is entirely created [2, 13, 14].

(iii) **Mutation Operator**

   In order to maintain the genetic diversity from one population generation to the next generation, the mutation operator is used. This solution may change entirely from the previous solution, and the best solution might be obtained by mutation operator [2, 13, 14].

Furthermore, as a modern evolutionary algorithm technique, GA provides significant advantages compared to traditional optimization algorithms. The most crucial advantage of GA is that it helps to prevent

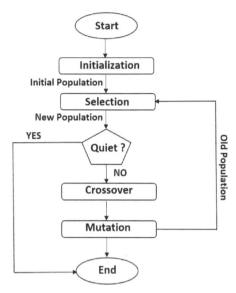

Figure 2.1 Flowchart of genetic algorithm [15].

getting stuck at local stationary points and increase the chance of finding the global optimum points. However, it is not ensured for finding the global optimum solution [2, 13, 14].

Simulated Annealing

Simulation annealing (SA) method, which is one of the most effective and general optimization algorithms of stochastic algorithms, is quite useful for finding the global minimum of a function of a considerable number of independent variables. Besides, the SA method includes the analogy between the physical annealing process and finding the minimum function value in mixed-integer, discrete, or continuous minimization problems. In condensed matter physics, the physical annealing process is known as a thermal process in order to obtain the low energy states of a solid in a heat bath. The fundamental notion of the SA algorithm is to use random search in terms of a Markov chain, which not only accepts changes that develop the objective function but also keeps some of the non-ideal changes.

At each iteration in the SA algorithm, a new point is randomly generated, and when any stopping criteria are satisfied, the algorithm ends (Figure 2.2). The distance between the new and current point or the extent of the search is based on Boltzmann probability distribution with a scale in proportion to the temperature. Boltzmann Probability Distribution [2, 16–18] is defined as

$$P(E) = e^{-E/kT}$$

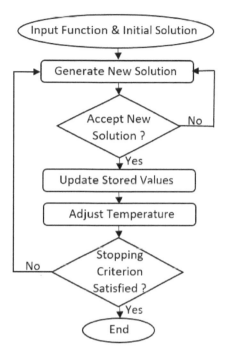

Figure 2.2 Flowchart of simulated annealing algorithm [16].

where,

P(E) : The Probability of Achieving the Energy Level (E),
k : The Constant of Boltzmann,
T : Temperature.

Differential Evolution

Differential Evolution (DE) is a search technique that was first introduced by Storn and Price in 1996 for optimization problems over continuous domains. DE is one of the most powerful real-parameter optimization algorithms at present. This algorithm comprises four basic stages: selection, crossover, mutation, and initialization. There exists three real control parameters in this algorithm: (i) differentiation/mutation constant, (ii) crossover constant and (iii) population size. The differential evolution performance relies on the manipulation of target and difference vector to obtain a trial vector. The other control parameters in DE algorithm are (i) problem dimension that scales the difficulty of the optimization case, and (ii) the maximum number of generations known as a stopping condition, and (iii) boundary constraints [2, 18]. A flowchart summarizing the process of DE algorithm is shown in Figure 2.3.

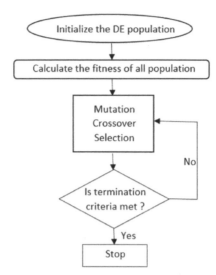

Figure 2.3 Flowchart of differential evolution algorithm [19].

The differential evolution algorithm is a population-based algorithm like GA that uses similar operators. The primary difference between these algorithms is that GA relies on a crossover, which is a mechanism of useful and probabilistic exchange of information among solutions in order to find better solutions. However, DE relies on mutation operation as the primary search mechanism. This main operation is based on the differences among randomly sampled pairs of solutions in the population. Even though this method is numerically uneconomical, DE is strong and efficient enough to find an optimum global value and to prevent the local minimum irrespective of initial points [20–22].

Particle Swarm Optimization

Particle Swarm Optimization (PSO) is a robust computational method inspired by the movement, foraging behaviors, and intelligence of swarms. This method uses the concept of simplified social interaction in order to optimize numerical problems iteratively [23, 24]. Comparison of the specific properties of PSO with other methods being more frequently used in optimization problems showed that the PSO method is more efficient than the others. In the optimization problems, this method is also indicated to have advanced computational skills [18].

In PSO, each individual searching for a solution is called a particle (agent), and the population of particles is called a swarm. Besides, the fitness function is used to understand how close an individual is to a solution. Each particle keeps track of its coordinates in the problem space, which is related to the best (fitness) solution achieved so far. Every iteration in PSO, every particle is updated by using two sequent best values. The first value is the best solution that has been reached up to that moment. This value is symbolized as P_{best}. Other best values being monitored by the particle swarm optimizer are the best values, achieved until that moment by any particle in the population. This best value is a global best represented as G_{best}.

In PSO, initially, the swarm and the necessary parameters are determined to search for the solution. The proximity function of the particles is measured with the help of the fitness function. The P_{best} and G_{best} values are updated according to these values. Then the velocity update of particles is calculated, and new positions are determined. The fitness function is used again to check how close it is to the solution. This cycle is repeated until the desired conditions are reached (see Figure 2.4). PSO concept includes changing the velocity of each particle against its 'P_{best}' and 'G_{best}' locations at each step. Acceleration is weighted by a random term with separate random numbers generated for acceleration against 'P_{best}' and 'G_{best}' locations. The updates of the particles are accomplished using the following Equations (2.1) and (2.2).

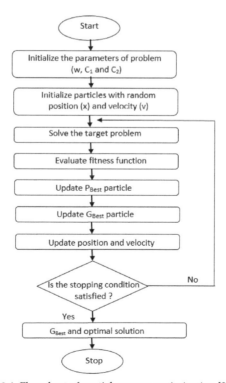

Figure 2.4 Flowchart of particle swarm optimization [2, 23, 24].

Equation (2.1) indicates a new velocity for each particle depending on the velocity of the previous particle, the best location ($\mathbf{P}_{best}$) achieved until that moment, and the global best location ($\mathbf{G}_{best}$) reached by the population. Equation (2.2) updates the individual position of the particle (X_i) in solution hyperspace.

In PSO velocity update equation,

$$V_{i+1} = w * V_i + C_1 * r_1 * (P_{best} - X_i) + C_2 * r_2 * (G_{best} - X_i) \tag{2.1}$$

$$X_{i+1} = X_i + V_{i+1} \tag{2.2}$$

where,

V_{i+1} = A new velocity for each particle based on its previous velocity

X_i = Updates individual particle's position

r_1, r_2 = The two random numbers

w = The constant parameter

V_i = The velocity term which indicates the correct dimension

Ant Colony Optimization

Ant Colony Optimization (ACO), which was developed by Dorigo and his collaborators in 1992, is a population-based metaheuristic algorithm that can be used to find approximate solutions for complex optimization problems. This algorithm was inspired by the ability of real ants to find the shortest path between nests and food points without their offensive visual hints and pheromone knowledge [18, 25].

A spectacular phenomenon which can be observed in nature is the marvelous trail followed by ants while foraging. Although the ants are blind, a queue is almost entirely formed by all the ants in the swarm. Scientists found that ants can achieve this by using a chemical composition called pheromone. Every pheromone of ants is used by the next ant who follows the same route by the probability of the track of these chemical hints. Ants prefer the shortest route. If the pheromone ratio increases in the route, that route is always preferred.

Nevertheless, the pheromone which is stored in the route evaporates in time. In the ant colony algorithm, this characteristic is remarked as three essential processes, which are known as state transition, local updating, and global updating. These categories organize together to construct a solution in a constructive approach.

In addition to these, ACO has a wide range of application areas to solve optimization problems such as traveling salesman, quadratic assignment, telecommunication, and water distribution network model, vehicle routing, and graph coloring. Especially, ACO is more useful for the traveling salesman and similar problems. In other words, ACO simulates

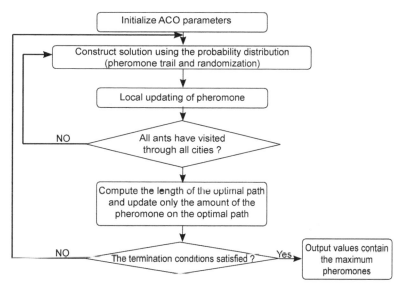

Figure 2.5 Flowchart of ant colony optimization [18, 25, 26].

the optimization of ant foraging behaviors, and the ACO procedure is demonstrated in Figure 2.5 [18, 25].

Artificial Bee Colony

Artificial Bee Colony (ABC) algorithm, which was first introduced by Karaboga in 2005, is an evolutionary algorithm based on the intelligent foraging behavior of honey bee colonies. The ABC algorithm, which is iterative, starts by associating all employed bees with randomly generated food solutions. The initial population of solutions is randomly filled with the number of food sources of dimensions generated [2].

The foraging process of bees inspires ABC algorithm in order to find out neat solutions to an optimization problem. Fundamental elements in ABC which were modeled after the foraging operations are indicated in below;

Fitness Value: Presents the profitability of a food source. It is presented as a single quantity related to an objective function of a proper solution.

Food Source: Presents a feasible solution to an optimization problem.

Bee Agents: A series of computational agents.

Karaboga and his research group have studied on ABC algorithm and its application in real life. They divided this algorithm into three groups: (i) scout bees, (ii) employed bees and (iii) onlooker bees. The colony is levelly divided into employed bees and onlooker bees. Each solution includes a set of optimization parameters, which symbolizes a location of food source in the search space. The number of employed bees is equal to the number of food sources. Each cycle of the search includes three steps: moving the employed and onlooker bees onto the food sources and calculating their nectar amounts, determining the scout bees and then moving them randomly onto the possible food sources.

In this algorithm, the points are assigned to be flowers, and the fitness value of the function at these points is equated with the nectar of the flower in the domain. Firstly, scout bees are randomly sent to evaluate selected points (flowers). The number of points evaluated is equal to the number of employed bees. The employed bees go to these points, calculate the relative fitness, and return to the hive with the nectar information. The onlooker bees look at the dance which is made by the employed bees and then choose a point to forage with a probability based on its fitness.

If the fitness value of a point in the neighborhood is better than the flower, the flower can be replaced by the point. The employed bees again go to forage. Until the optimum value is obtained, the cycle of the onlooker and employed bees is repeated. If the fitness of any flower does not change for a certain number of cycles, it is considered to be infertile and is ejected. The employed bees of these flowers turn into scouts and randomly choose another source [2, 27–30]. The flowchart of the ABC algorithm is presented in Figure 2.6.

Markov Chain Monte Carlo

Markov Chain Monte Carlo (MCMC) technique, which is a computer-driven sampling method to obtain information about distributions, and their use in data modeling for numerical integration. This method can be used for both the stochastic and deterministic problems. As a global numerical technique, the MCMC method became feasible with the development of computers, and its application maintained to expand with every reproduced computer generation. The method permits for the symbolization of a distribution without being informed all of the properties of mathematical distribution by randomly sampling values out of the distribution.

The absolute power of MCMC is that it can be used to draw samples from distributions even when all of them and what is known about the distribution is how to evaluate the density for different sampling. Also, the properties of a distribution can be estimated by examining random samples from the distribution using the Monte Carlo method. For instance, rather than finding the average of a normal distribution by calculating it from the equation distribution, a Monte–Carlo technique would be to draw a considerable amount of random samples from a normal distribution and

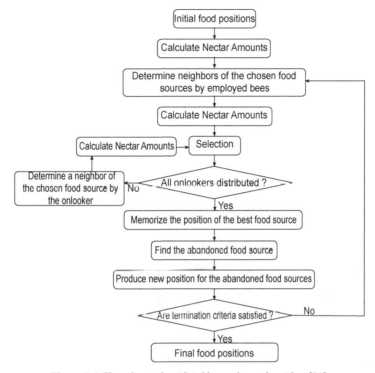

Figure 2.6 Flowchart of artificial bee colony algorithm [31].

calculate the sample mean. That is, the Monte Carlo technique calculates the mean of a large sample of numbers rather than calculate the mean directly from the standard equation distributions. This advantage of MCMC is most apparent if random samples are simple to draw, and the equation distributions are hard to work by other means.

Nevertheless, there are some drawbacks to the MCMC method. For instance, it is time-consuming because there is a need to generate a wide range of sampling to obtain the desired output, and the results in the method are not exact, only the approximation of values.

MCMC is a standard method to obtain information about distribution, in particular, forecasting posterior distribution for the Bayesian inference. In this situation, MCMC permits the user to approximate aspects of posterior distributions that cannot be calculated directly [32, 33].

Conclusion

This chapter presents the review of the seven most preferred stochastic optimization methods in detail for use in different industrial areas such as engineering, construction, automotive, textile, and biomedical. These include Genetic Algorithm (GA), Simulated Annealing (SA), Differential Evolution (DE), Particle Swarm Optimization (PSO), Ant Colony Optimization (ACO), Artificial Bee Colony (ABC) and Markov Chain Monte Carlo (MCMC). A brief description of each method is presented along with their flowcharts. All the stochastic methods discussed can be used for convenience to solve the several problems and comparable results with a high accuracy can be obtained. This chapter gives the general mathematical foundations and algorithmic frameworks of these seven methods, and gives us insights about how these optimization algorithms can be effectively used in such a wide range of application areas.

References

[1] Ombach, J. 2014. A short introduction to stochastic optimization. Schedae Informaticae. 23: 9–20.
[2] Rao, R.V. and Savsani, V.J. 2012. Mechanical Design Optimization using Advanced Optimization Techniques. Springer. London.
[3] Tech Target, whatis.techtarget.com/definition/stochastic-optimization.
[4] Yan, X., Zhou, Y., Wen, Y. and Chai, X. 2013. Qualitative and quantitative integrated modeling for stochastic simulation and optimization. J. Appl. Math. Volume 2013, Article ID 831273.1–12.
[5] Li, G. and Zhang, W. 2013. Study on Indefinite stochastic linear quadratic optimal control with inequality constraint. J. Appl. Math. Volume 2013, Article ID 805829, 1–9.
[6] Aydin, L., Aydin, O., Artem, H.S. and Mert, A. 2016. Design of dimensionally stable composites using efficient global optimization method. J. Mater. Des. Appl. 1–13.
[7] Zakaria, A., Firas, B.I., Hossain-Lipu, M.S. and Hannan, M.A. 2019. Uncertainty models for stochastic optimization in renewable energy applications. Renew. Energy. 145: 1543–1571.
[8] Niamsup, P. and Rajchakit, G. 2013. New results on robust stability and stabilization of linear discrete-time stochastic systems with convex polytopic uncertainties. J. Appl. Math. Vol. 2013, Article ID 368259.1–10.
[9] Maggioni, F., Cagnolari, M., Bertazzi, L. and Wallace, S.W. 2019. Stochastic optimization models for a bike-sharing problem with transshipment. Eur. J. Oper. Res. 276(1): 272–283.
[10] Gutierrez, G., Galan, A., Sarabia, A. and De-Prada, C. 2018. Two-stage stochastic optimization of a hydrogen network. IFAC 18: 263–268.
[11] Khayyam, H., Naebe, M., Bab-Hadiashar, A., Jamshidi, F., Li, Q., Atkiss, S., Buckmaster, D. and Fox, B. 2015. Stochastic optimization models for energy management in carbonization process of carbon fiber production. Appl. Energy. 158: 643–655.
[12] Tifkitsis, K.I., Mesogitis, T.S., Struzziero, G. and Skordos, A.A. 2018. Stochastic multi-objective optimisation of the cure process of thick laminates. Compos. Part A 112: 383–394.
[13] www.geeksforgeeks.org/genetic-algorithms.
[14] Aydin, L. and Artem, H.S. 2011. Comparison of stochastic search optimization algorithms for the laminated composites under mechanical and hygrothermal loadings. J. Reinf. Plast. Comp. 30(14): 1197–1212.
[15] Dastanpour, A. and Mahmood, R.A.J. 2013. Feature selection based on genetic algorithm and support vector machine for intrusion detection system. SDIWC 169–181.

[16] Ozturk, S., Aydin, L. and Celik, E. 2018. A comprehensive study on slicing processes optimization of silicon ingot for photovoltaic applications. J. Sol. Energy. 161: 109–124.

[17] Grabusts, P., Musatovs, J. and Golenkov, V. 2018. The application of simulated annealing method for optimal route detection between objects. Procedia Comput. Sci. 149: 95–101.

[18] Aydin, L. and Artem, H.S. 2017. Design and optimization of fiber composites, Turkey. pp. 299–315. *In*: Seydibeyoglu, M.O., Mohanty, A.K. and Misra, M. [eds.]. Woodhead Publishing Series in Composites Science and Engineering.

[19] Huang, Z. and Chen, Y. 2013. An improved differential evolution algorithm based on adaptive parameter. Journal of Control Science and Engineering Volume 2013, Article ID 462706, 1–5.

[20] Das, S. and Suganthan, P.N. 2011. Differential evolution: A survey of the state-of-the-art. IEEE T. Evolut. Comput. 15(1): 4–35.

[21] Jujjavarapu, S. and Chimmiri, V. 2016. Dynamic modeling and metabolic flux analysis for optimized production of rhamnolipids. Chem. Eng. Commun. 203: 326–338.

[22] Arunachalam, V. 2008. Optimization using Differential Evolution. M.S. Thesis, The University of Western Ontario, London.

[23] Koyuncu, H. and Ceylan, R. 2019. A PSO based approach: Scout particle swarm algorithm for continuous global optimization problems. J. Comput. Des. Eng. 6: 129–142.

[24] Jammani, J.G. and Pandya, M. 2019. Coordination of SVC and TCSC for Management of Power Flow by Particle Swarm Optimization. 5th International Conference on Power and Energy Systems Energy Procedia. Nagoya, Japan 156: 321–326.

[25] Ali, H. and Kar, A.K. 2018. Discriminant analysis using ant colony optimization—an intra-algorithm exploration. International Conference on Computational Intelligence and Data Science 132: 880–889.

[26] Kaveh, A. and Talatahari, S. 2009. Particle swarm optimizer, ant colony strategy and harmony search scheme hybridized for optimization of truss structures. Comput. Struct. Journal 87: 267–283.

[27] Bhokray, K. 2013. Artificial Bee Colony Optimization. Indian Institute of Technology, Bombay.

[28] Scaria, A., George, K. and Sebastian, J. 2016. An artificial bee colony approach for multi-objective job shop scheduling. Procedia Eng. 25: 1030–1037.

[29] Janakiraman, S. 2018. A hybrid ant colony and artificial bee colony optimization algorithm based cluster head selection for IoT. 8th International Conference on Advances in Computing and Communication. Procedia Computer Science. India 143: 360–366.

[30] Anuar, S., Selamat, A. and Sallehuddin, R. 2016. A modified scout bee for artificial bee colony algorithm and its performance on optimization problems. J. King Saud Univ., Comp. & Info. Sci. 28: 395–406.

[31] Mohamed, A.F., Elarini, M.M. and Othman, A.M. 2014. A new technique based on Artificial Bee Colony Algorithm for optimal sizing of stand-alone photovoltaic system. J. Adv. Res. 5: 397–408.

[32] Ravenzwaaij, D., Cassey, P. and Brown, S.D. 2018. A simple introduction to Markov Chain Monte–Carlo sampling. Psychonomic Bulletin & Review 25(1): 143–154.

[33] Wang, Z. 2019. Markov chain Monte Carlo sampling using a reservoir method. Computational Statistics & Data Analysis 139: 64–74.

CHAPTER 3

Mathematica and Optimization

Melih Savran,[1] *Harun Sayi*[2] *and Levent Aydin*[3,*]

Global and Local Optimization by Mathematica

The Mathematica software has a collection of commands which make exact-numeric optimization to solve linear-nonlinear and unconstrained-constrained problems. In this respect, *NMinimize* and *NMaximize* commands are used in numeric global optimization methods while *Minimize* and *Maximize* commands are only appropriate for exact global optimization. Numeric local optimization is carried out by using the *FindMinimum* command. The abovementioned commands could all be utilized for linear-nonlinear and constrained-unconstrained optimization problems [1]. Detail explanations about commands, algorithms, and which types of problems they are used to solve are given in Table 3.1 and Figure 3.1.

Numerical global optimization algorithms for constrained nonlinear problems can be widely classified into gradient-based methods and direct search methods. Gradient-based methods make use of first or second derivatives of objective function and constraints for calculation, while Direct search methods have a probabilistic process and do not need derivative information.

In this chapter, the Mathematica commands (*FindMinimum, NMaximize, and Nminimize, RandomSearch, SimulatedAnnealing,*

[1] İzmir Katip Çelebi University, Department of Mechanical Engineering, Izmir, Turkey.
  Email: mlhsvrn@gmail.com
[2] İzmir Institute of Technology, Department of Mechanical Engineering, Izmir, Turkey.
  Email: harunsayi07@gmail.com
[3] İzmir Kâtip Çelebi University, Department of Mechanical Engineering, Izmir, Turkey.
* Corresponding author: leventaydinn@gmail.com

Table 3.1 Optimization methods and commands [1].

Optimization Types	Optimization Methods/Algorithms	Mathematica Commands
• Numerical Local Optimization	• Linear Programming Methods • Nonlinear Interior Point Algorithms	• *FindMinimum* • *FindMaximum*
• Numerical Global Optimization	• Linear Programming Methods • Differential Evolution • Nelder-Mead • Simulated Annealing • Random Search	• *NMinimize* • *NMaximize*
• Exact Global Optimization	• Linear Programming Methods, • Cylindrical Algebraic Decomposition • Lagrange Multipliers • Integer Linear Programming	• *Minimize* • *Maximize*
• Linear Optimization	• Linear Programming Methods (simplex, revised simplex, interior point)	• *LinearProgramming*

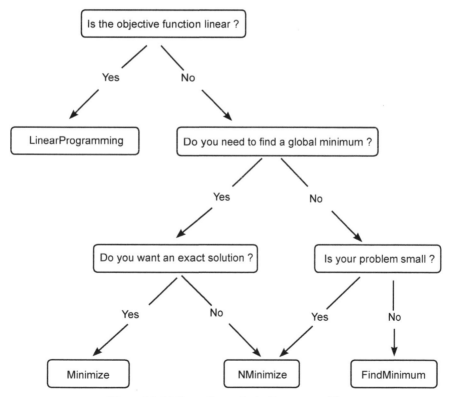

Figure 3.1 Mathematica optimization process [1].

NelderMead, DifferentialEvolution) are explained and the capability of the algorithms are evaluated for finding the global minimum for distinct test functions.

Random Search

The Random Search (RS) algorithm implemented by Mathematica has a stochastic approach. In the working process, the algorithm composes population, including random starting points, and then the algorithm evaluates the convergence behavior of the starting points to the local minimum utilizing the *FindMinimum* local search method. During this process, the options: (i) *SearchPoints* determines the number of starting points as per "min(10 f,100)" expression, where f is the number of variables, (ii) *RandomSeed* adjusts the starting value for random number producer, (iii) Method is defined by which method to use for minimizing the objective function by *FindMinimum*. In here, for unconstrained optimization problems, *FindMinimum* command uses Quasi-Newton as search method which does not need the second derivatives (Hessians matrix) to be computed; instead, the Hessian is updated by analyzing successive gradient vectors. In the case of the constrained optimization problem, the non-linear interior point is selected as a search method by the *FindMinimum* command, (iv) *PostProcess* option can be selected as Karush–Kuhn–Tucker (KKT) conditions or *FindMinimum*. At the end of these processes, the best local minimum is selected to be the solution.

Mathematica automatically controls the options *InitialPoints, Method, PenaltyFunction, PostProcess,* and *SearchPoints* used in random search algorithm, and appropriate values of options are selected according to optimization problems [1]. The RS algorithm follows the procedure given in Figure 3.2.

In order to evaluate the performance capacity of the Random Search algorithm in finding the global minimum, separable and non-separable multimodal test functions having more than one, few or many local minima are used. This kind of global optimization problems are quite hard when an algorithm is not designed appropriately, and it can be inserted into the local minima without finding the global minimums or not all global minimums. In this respect, the first selected test function, which has global minima is located at $f(0,0) = 0$ is Ackley [3]. The following commands give the Mathematica syntax for the definition of Ackley function and its 3D plot in an interval.

```
In[1]:=  f[x1_,x2_]:=-20Exp[(-0.02Sqrt[0.5(x1^2+x2^2)])]
         -Exp[(0.5(Cos[2Pix1]+Cos[2Pix2]))]+20+Exp[1];
In[2]=   Plot3D[f[x1,x2],{x1,-35,35},{x2,-35,35},
         AxesLabel->{x1,x2,y}]
```

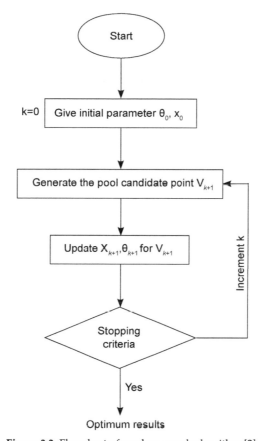

Figure 3.2 Flowchart of random search algorithm [2].

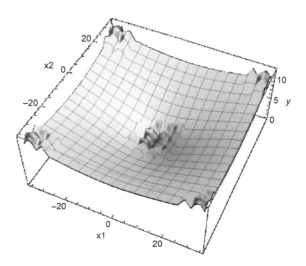

It is noted that the **RandomSearch** command may not find a global minimum without working any alteration of its options.

```
In[3]:=   NMinimize[{f[x1,x2],-35≤x1≤35,-35≤x2≤35},
          {x1,x2},Method->"RandomSearch"]
Out[3]=   {2.83635, {x1->-5.99749,x2->8.99623}}
```

Sometimes changing the search point option that specifies the number of points to start searches can be effective in finding a global minimum.

```
In[4]:=   Do[Print[NMinimize[{f[x1,x2],-35≤x1≤35,-35≤x2≤35},
          {x1,x2},Method->{"RandomSearch","SearchPoints"->i}]],
          {i,500,3000,500}]

          {0.39531,{x1->0.996345,x2->0.996345}}
          {0.280127,{x1->-5.04225*10^-24,x2->-0.9948}}
          {0.280127,{x1->-5.04225*10^-24,x2->-0.9948}}
          {0.280127,{x1->-5.04225*10^-24,x2->-0.9948}}
          {0.280127,{x1->-5.04225*10^-24,x2->-0.9948}}
          {1.2012*10^-9,{x1->-8.42728*10^-
          10,x2->-4.16243*10^-9}}
```

The effect of the **RandomSeed** option, which constitutes starting value for the random number generator, can be investigated below. In the previous example, while the *"Searchpoints"->500* is not sufficient to reach the global minimum, in the following example a global minimum can be obtained by setting the values of the *SearchPoints* and the *RandomSeed* to 500 and 5, respectively.

```
In[5]:=   Do[Print[NMinimize[{f[x1,x2],-35≤x1≤35,-35≤x2≤35},
          {x1,x2},Method->{"RandomSearch","SearchPoints"->500,
          "RandomSeed"->i}]],{i,5}]

          {0.280127,{x1->-7.38323*10^-25,x2->0.9948}}
          {7.40815*10^-10,{x1->6.89861*10^-10,
          x2->-2.52669*10^-9}}
          {0.280127,{x1->5.59478*10^-24,x2->0.9948}}
          {0.39531,{x1->0.996345,x2->0.996345}}
          {1.37499*10^-9,{x1->-3.64123*10^-9,
          x2->-3.22083*10^-9}}
```

In here, points are produced on a grid to utilize as initial points. If the approximate solution range of the problem can be estimated, assigning the starting point makes it easier to get the solution.

```
In[6]:=   Print[NMinimize[{f[x1,x2],-35≤x1≤35,-35≤x2≤35},
          {x1,x2},Method->{"RandomSearch","InitialPoints"-
          >Flatten[Table[{i,j},{i,-35,35,5},{j,-35,35,5}],1]}]]
Out[6]=   {-4.44089*10^-16,{x1->-1.52703*10^-15,
          x2->-1.52703*10^-15}}
```

PostProcess option is not of primary importance for this problem. *PostProcess* methods *KKT* and *FindMinimum* give the same results.

```
In[7]:=   Print[NMinimize[{f[x1,x2],-35≤x1≤35,-35≤x2≤35},
          {x1,x2},Method->{"RandomSearch", "SearchPoints"->3000,
          "PostProcess"->KKT}]]

Out[7]=   {1.2012*10^-9,{x1->-8.42726*10^-10,
          x2->-4.16243*10^-9}}

In[8]:=   Print[NMinimize[{f[x1,x2],-35≤x1≤35,-35≤x2≤35},
          {x1,x2},Method->{"RandomSearch", "SearchPoints"->3000,
          "PostProcess"->FindMinimum}]]

Out[8]=   {1.2012*10^-9,{x1->-8.42728*10^-10,
          x2->-4.16243*10^-9}}
```

Another test function Holder Table 1, which is separable and multimodal, is used to evaluate the capability of *RandomSearch* command in finding the global minimum [3]. This test function has global minima in located at f (±9.646168, ±9.6461680) = −26.920336. The followings are Mathematica syntax for the definition of the "Holder Table 1" function and its 3D plot.

```
Clear[f];
In[9]:= f[x1_,x2_]:=-Abs[Cos[x1]Cos[x2]Exp
[Abs[1-((x1^2+x2^2)^0.5)/Pi]]];
In[10]:= Plot3D[f[x1,x2],{x1,-10,10},{x2,-10,10}]
```

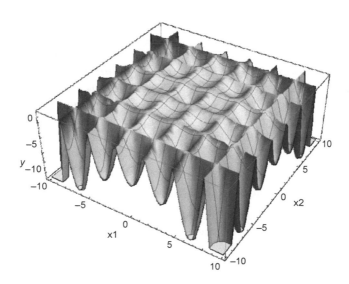

The RS algorithm finds one of the global minimum points without working any alteration of its options for this problem.

```
In[11]:=  NMinimize[{f[x1,x2],-10≤x1≤10,-10≤x2≤10},
          {x1,x2},Method->"RandomSearch"]

Out[11]=  {-26.9203, {x1 -> -9.64617, x2 -> -9.64617}}

In[12]:=  Do[Print[NMinimize[{f[x1,x2],-10≤x1≤10,-10≤x2≤10},
          {x1,x2},Method->{"RandomSearch", "RandomSeed"->i}]],
          {i,{1,6,7}}]

          {-26.9203,{x1->-9.64617,x2->9.64617}}
          {-26.9203,{x1->-9.64617,x2->-9.64617}}
          {-26.9203,{x1->9.64617,x2->9.64617}}
```

Simulated Annealing

The Simulated Annealing (SA) algorithm implemented by Mathematica is a stochastic approach having a working process based on the physical annealing procedure of solids. The SA is designed to find the largest or smallest values of functions having many variables and the smallest values of nonlinear functions having many local minimums. The algorithm is named Simulated Annealing because it exemplifies the perfect arrangement of atoms of solid bodies and minimizing the potential energy during the cooling process. The algorithm allows the structure to move away from the local minimum and to investigate and to find a better global minimum [4].

In the working process for each iteration; firstly, the startup solution "Z" is produced, Secondly, "Z_{new}" is generated in the neighborhood of the current point, "Z" and then Z_{best} is defined.

If $f(Z_{new}) \leq f(Z_{best})$, Z_{new} replaces Z_{best} and Z. Otherwise, Z_{new} replaces with Z. In this loop, options *InitialPoints, SearchPoints,* and *RandomSeed* are capable of determining the initial guess and its number and starting value, respectively. In the SA algorithm, random movements in the search space are performed based on the Boltzmann probability distribution $e^{D(k, \Delta f, f_0)}$. Here D is the function defined by option *BoltzmannExponent*, k is the current iteration, Δf is the change in the objective function. In the Mathematica, if the user does not select manually, B is defined as $\frac{-\Delta f log(k+1)}{10}$ by *BoltzmannExponent*.

For all starting points, the working process introduced above is returned by the time either the algorithm converges to a point, or the algorithm remains at the same point as a result of the number of iterations assigned by the option *LevelIterations* [5]. The SA algorithm follows the procedure given in Figure 3.3.

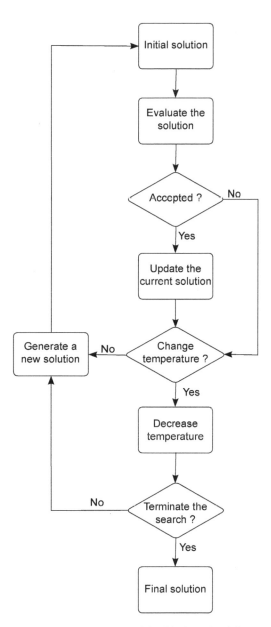

Figure 3.3 Flowchart of the SA algorithm [6].

"Ackley" and "Holder Table 1" is used to evaluate the performance capacity of the *SimulatedAnnealing* command to find the global minimum.

```
In[1]:=   f[x1_,x2_]:=-20 Exp[(-0.02 Sqrt[0.5 (x1^2+x2^2)])]-
          Exp[(0.5 (Cos[2 Pi x1]+Cos[2 Pi x2]))]+20+Exp[1];
In[2]=    Plot3D[f[x1,x2],{x1,-35,35},{x2,-35,35},
          AxesLabel->{x1,x2,y}]
```

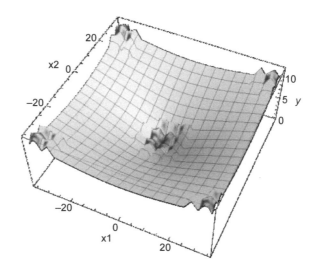

The SA algorithm may not find a global minimum by using the default value of its options.

```
In[3]:=   NMinimize[{f[x1,x2], -35≤x1≤35,-35≤x2≤35},
          {x1,x2},Method->{"SimulatedAnnealing"}]
Out[3]=   {2.37578, {x1 -> 7.99584, x2 -> 3.99792}}
```

BoltzmannExponent includes a function which determines a new point at each iteration; thus, the *BoltzmannExponent* is a significant option that shows the way to achieve a global minimum. If this function is utilized without a default value, the obtained result can be changed. However, in the following problem, changing this option alone has not been enough to find the global minimum.

```
In[4]:=   NMinimize[{f[x1,x2],-35≤x1≤35,-35≤x2≤35},{x1,x2},
          Method->{"SimulatedAnnealing", "BoltzmannExponent"
          ->Function[{i,df,f0},-df/(Exp[i/10])]}]
Out[4]=   {0.830095, {x1 -> -2.99495, x2 -> 6.41153*10^-9}}
```

For this problem, although the *PerturbationScale* alters the result, changing this option alone has not been enough to find the global minimum. The algorithm attaints to local minimum points.

```
In[5]:=   Do[Print[NMinimize[{f[x1, x2], -35 ≤ x1 ≤ 35, -35 ≤ x2
          ≤ 35}, {x1, x2}, Method -> {"SimulatedAnnealing",
          "PerturbationScale" -> i}]], {i, 15}]
```

```
{2.37578,{x1->7.99584,x2->3.99792}}
{2.40345,{x1->0.999488,x2->8.99539}}
{1.0993,{x1->-1.04986*10^-9,x2->3.99502}}
{3.8527,{x1->-1.99944,x2->14.9958}}
{6.15308,{x1->-23.9966,x2->-9.9986}}
{4.50046,{x1->14.9966,x2->-9.99773}}
{4.26698,{x1->11.9971,x2->-11.9971}}
{4.27353,{x1->7.99805,x2->-14.9963}}
{2.63697,{x1->5.99725,x2->-7.99634}}
{6.15308,{x1->-23.9966,x2->-9.9986}}
{6.15308,{x1->-23.9966,x2->-9.9986}}
{6.15308,{x1->-23.9966,x2->-9.9986}}
{6.15308,{x1->-23.9966,x2->-9.9986}}
{6.15308,{x1->-23.9966,x2->-9.9986}}
{6.15308,{x1->-23.9966,x2->-9.9986}}
```

Using many more *SearchPoints,* a global minimum can be obtained.

```
In[6]:=   Do[Print[NMinimize[{f[x1,x2],-35≤x1≤35,-35≤x2≤35},
          {x1,x2},Method-> {"SimulatedAnnealing",
          "SearchPoints"->i}]],{i,100,500,100}]
```

```
{0.830095,{x1->-2.99495,x2->7.32049*10^-10}}
{0.62186,{x1->1.99543,x2->-0.997715}}
{0.280127,{x1->-1.64485*10^-9,x2->-0.9948}}
{0.280127,{x1->0.9948,x2->5.25186*10^-12}}
{1.937*10^-9,{x1->-2.31279*10^-9,x2->-6.44598*10^-9}}
```

As previously seen, while changing the search points alone is sufficient to find the global minimum, in the case of conducting a search using the *RandomSeed, PerturbationScale,* and *BoltzmannExponent,* the algorithm seizes the local minimums.

```
In[7]:=   Do[Print[NMinimize[{f[x1,x2],-35≤x1≤35,
          -35≤x2≤35},{x1,x2},Method->
          {"SimulatedAnnealing","RandomSeed"->i}]],{i,0,10}]
```

```
{2.37578,{x1->7.99584,x2->3.99792}}
{0.557056,{x1->-4.99634*10^-9,x2->1.99487}}
{2.15456,{x1->7.99533,x2->-0.999416}}
{0.39531,{x1->0.996345,x2->0.996345}}
{3.46466,{x1->-8.99708,x2->9.99676}}
{0.993567,{x1->2.99583,x2->1.99722}}
{1.58244,{x1->-2.9975,x2->-4.99584}}
{1.22508,{x1->-3.99557,x2->1.99779}}
{1.46596,{x1->1.99819,x2->-4.99546}}
{0.39531,{x1->-0.996345,x2->0.996345}}
{2.29034,{x1->4.99729,x2->6.9962}}
```

```
In[8]:=  Do[Print[NMinimize[{f[x1,x2],-35≤x1≤35,-35≤x2≤35},
         {x1,x2},Method-> {"SimulatedAnnealing",
         "PerturbationScale"->3,"SearchPoints"->500,
         "RandomSeed"->i}]], {i, 0, 10, 1}]

         {-4.44089*10^-16,{x1->-1.62365*10^-15,
         x2->2.19073*10^-16}}
         {0.39531,{x1->-0.996345,x2->0.996345}}
         {0.557056,{x1->1.99487,x2->-1.44602*10^-11}}
         {1.16405,{x1->-2.99649,x2->-2.99649}}
         {0.557056,{x1->-1.99487,x2->-6.2523*10^-11}}
         {0.62186,{x1->-1.99543,x2->-0.997715}}
         {0.557056,{x1->1.99487,x2->7.97744*10^-12}}
         {0.39531,{x1->0.996345,x2->0.996345}}
         {0.280127,{x1->0.9948,x2->-6.58993*10^-9}}
         {2.09443*10^-9,{x1->-7.39447*10^-9,
         x2->-3.93044*10^-10}}
         {0.993567,{x1->-1.99722,x2->-2.99583}}
```

```
Clear[f]
In[9]:= f[x1_,x2_]:=-Abs[Cos[x1]Cos[x2]
Exp[Abs[1-((x1^2+x2^2)^0.5)/Pi]]];
In[10]:= Plot3D[f[x1,x2],{x1,-10,10},{x2,-10,10}]
```

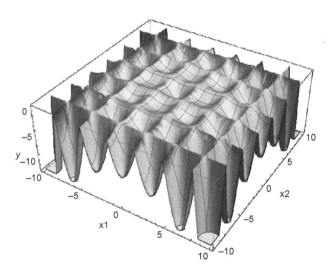

The Simulated Annealing algorithm finds one of the global minimum points without working any alteration of its options for this problem.

```
In[11]:=  NMinimize[{f[x1,x2],-10≤x1≤10,-10≤x2≤10},{x1,x2},
          Method->"SimulatedAnnealing"]
Out[11]=  {-26.9203, {x1 -> 9.64617, x2 -> 9.64617}}
```

Unlike the Random Search algorithm, four distinct global minimum points can be found by using the Simulated Annealing algorithm.

```
In[12]:= Do[Print[NMinimize[{f[x1,x2],-10≤x1≤10,
         -10≤x2≤10},{x1,x2},Method->
         {"SimulatedAnnealing","RandomSeed"->i}]],
         {i,{1,2,3,11}}]
         {-26.9203,{x1->9.64617,x2->9.64617}}
         {-26.9203,{x1->-9.64617,x2->-9.64617}}
         {-26.9203,{x1->-9.64617,x2->9.64617}}
         {-26.9203,{x1->9.64617,x2->-9.64617}}
```

FindMinimum

The FindMinimum command is used to find a local minimum function for unconstrained and constrained optimization problems [1].

The options of the *FindMinimum* command are *Method, MaxIterations, WorkingPrecision, PrecisionGoal*, and *AccuracyGoal.*

The *Method* option ventilates that the *FindMinimum* command selects which method solves problems. Hereof, unconstrained optimization problems; (i) *Newton* utilizes the exact Hessian or a finite difference approximation, (ii) *QuasiNewton* uses the quasi-Newton BFGS approximation which was composed by updates based on past steps, (iii) the *LevenbergMarquardt* method also known as the damped least-squares (DLS) method, is employed to solve non-linear least-squares problems, (iv) the *ConjugateGradient* method is appropriate for solving linear systems, (v) the *PrincipalAxis* method does not need derivatives and it requires two starting conditions in each variable. In constrained optimization problems, only *InteriorPoint* can be selected as a method.

The *MaxIterations* option indicates the maximum number of iterations which ought to be utilized. In the constrained optimization problems, the default *"MaxIterations"*->500 is used.

WorkingPrecision, PrecisionGoal, and *AccuracyGoal* are options specifying the number of digits of precision. The former controls the internal computations while the latter checks the final result. By default, *WorkingPrecision*->**prec** is equal to *MachinePrecision* but If **prec**>*MachinePrecision* a constant **prec** value which is used during the computation. When *AccuracyGoal* and *PrecisionGoal* options are selected as *Automatic*, the default values are set to *WorkingPrecision*/3 and *Infinity*, respectively [1].

Carrom table function, which is a non-separable and multimodal function, and has many local minima has taken as a test function, and the *FindMinimum* command and the effect of its options in finding the local minima are investigated [3].

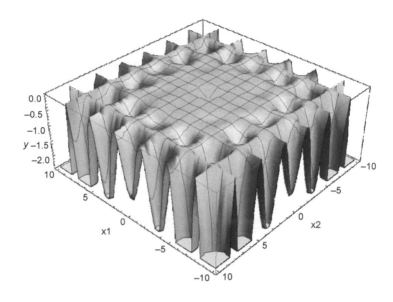

```
In[1]:=   f[x1_,x2_]:=-(Cos[x1]Cos[x2]
          Exp[Abs[1-((x1^2+x2^2)^0.5)/Pi]])^2/30
In[2]:=   Plot3D[f[x1,x2],{x1,-10,10},{x2,-10,10},
          AxesLabel->{x1,x2,y}]
In[3]:=   FindMinimum[{f[x1,x2],-10≤x1≤10,-10≤x2≤10},{x1,x2}]
Out[3]=   {-24.1568,{x1->9.64617,x2->9.64617}}
In[4]:=   FindMinimum[{f[x1,x2],-10≤x1≤10,-10≤x2≤10},
          {x1,x2},Method->"InteriorPoint"]
Out[4]=   {-0.246302, {x1 -> -1.22418*10^-14,
          x2 -> -1.29143*10^-14}}
In[5]:=   Do[Print[FindMinimum[{f[x1,x2],-10≤x1≤10,
          -10≤x2≤10},{x1,x2},Method->"InteriorPoint",
          "MaxIterations"->i]],
          {i,{1,10,100,500,1000,2000,4000,8000}}]

          {-0.0105322,{x1->0.969586,x2->0.969586}}
          {-0.246302,{x1->-8.74067*10^-8,x2->-8.74067*10^-8}}
          {-0.246302,{x1->-8.37899*10^-15,x2->-8.38925*10^-15}}
          {-0.246302,{x1->-1.22418*10^-14,x2->-1.29143*10^-14}}
          {-0.246302,{x1->-1.22418*10^-14,x2->-1.29143*10^-14}}
          {-0.246302,{x1->-1.22418*10^-14,x2->-1.29143*10^-14}}
          {-0.246302,{x1->-1.22418*10^-14,x2->-1.29143*10^-14}}
          {-0.246302,{x1->-1.22418*10^-14,x2->-1.29143*10^-14}}
```

```
In[6]:=    Table[Print[FindMinimum[{f[x1,x2],-10≤x1≤10,
           -10≤x2≤10},{{x1,RandomReal[{-10,10}]},
           {x2,RandomReal[{-10,10}]}},Method->
           "InteriorPoint"]],{10}]

           {-0.0368271,{x1->0.000019185,x2->-3.44978}}
           {-1.42781,{x1->6.50458,x2->-6.50458}}
           {-6.7549,{x1->9.68366,x2->-6.45799}}
           {-0.272117,{x1->3.63079*10^-7,x2->-6.59135}}
           {-1.42781,{x1->6.50458,x2->-6.50458}}
           {-2.01069,{x1->-1.67999*10^-7,x2->9.73295}}
           {-1.42781,{x1->-6.50458,x2->-6.50458}}
           {-0.436543,{x1->6.56051,x2->3.28309}}
           {-0.0843916,{x1->-3.36299,x2->3.36298}}
           {-2.78243,{x1->-9.71802,x2->3.24199}}
```

NMinimize and NMaximize Functions

These functions in Mathematica enable us to optimize complex problems in science and engineering and their certain characteristic features using search algorithms. Although these methods are efficient in finding global optima, it might be difficult to find optimum results even without constraints and boundary conditions. The best way to cope with this situation might be optimizing given functions with different initial conditions. The following examples are obtained again using the initial test functions; Ackley function of f (x1, x2) and Holder Table 1 function of g(x3, x4), respectively.

```
In[15]:=   NMinimize[{f[x1,x2],-35≤x1≤35,-35≤x2≤35},{x1,x2}]
Out[15]=   {0.8740,{x1→-0.9984,x2→-2.9952}}
In[4]:=    NMaximize[{f[x1,x2],-35≤x1≤35,-35≤x2≤35},{x1,x2}]
Out[4]=    {12.3202,{x1→34.5137,x2→34.51377}}
In[7]:=    NMinimize[{g[x3,x4],-10≤x3≤10,-10≤x4≤10},{x3,x4}]
Out[7]=    {-26.9203,{x3→9.6461,x4→9.6461}}
In[8]:=    NMaximize[{g[x3,x4],-10≤x3≤10,-10≤x4≤10},{x3,x4}]
Out[8]=    {-2.5326×[10]^(-13),{x3→-4.7498,x4→-4.7123}}
```

According to initial results, global minima and maximal values of Ackley Function might be achived. However, it is seen that it was not valid for Holder Table 1 function. Adjustment of the parameters or changing the restriction region might be effective towards obtaining global values.

Constraints might be either in the list form or a rational combination of domain options, equalities, and inequalities. For example, if one needs to specify results in integer form, the unknown parameter **z** should be included as z ∈ *Integers* in line. Then, this constraint restricts the possible solutions as being only integers. Besides, the **NMinimize** command requires a rectangular starting region to begin optimization. It means that each variable in the given function should have a finite upper and lower bound. Using the **Method** option enabling us to apply other types of search algorithms is a way to provide non-automatic set solutions as seen previous parts of this chapter performed using *SA* and *RS* algorithms. Here, it can be said that if the function being minimized or maximized (called as an objective function) and constraints are linear, the **LinearProgramming** method is the default setting in the solving process. If the central part of the objective function is not numerical and also the variables are in integer form, *DE* is the algorithm as default. In other situations, *NM* is the search algorithm to be used. If **NelderMead** does not provide desirable solutions, it switches with *DE* to obtain optimum values [1].

Differential Evolution

Differential evolution (DE) is one of the most common stochastic search algorithms in the optimization and solution of complicated and challenging design problems. As mentioned in the previous chapter of this book (Chapter 2), the algorithm is built on four main steps which are initialization, mutation, crossover, and selection. Although *DE* is an efficient search algorithm thanks to covering a population of solutions in iterations rather than a single solution, it computationally requires more process time which makes it an expensive method. DE is a robust and reliable algorithm to obtain global optimum. However, there is uncertainty finding global optimum points as also valid for other types of search methods [7].

In iterations, a new population of k points is produced. Then, the jth new point is produced by taking three random points such as $z_1, z_2,$ and z_3 from the previously generated population. Then it builds the new formation by $z_s = z_3 + s(z_1 - z_2)$ that s is the real scaling factor. A new point z_{new} is generated from z_j and z_s by picking the ith coordinate or another coordinate of jth from z_s with the probability of p. Then, z_{new} changes with z_j, if the function of $h(z_{new})$ is smaller than the function of $h(z_j)$ [1].

The command **DifferentialEvolution** consists of specific adjustment options which are **CrossProbability (P), InitialPoints, PenaltyFunction, PostProcess, RandomSeed, ScalingFactor, SearchPoints,** and **Tolerance** whether none of them does guarantee to find global optima. The process flow chart of the algorithm is illustrated in Figure 3.4 [8].

As performed with previous search algorithms, the same test functions of Ackley and Holder Table 1 are used to evaluate the capacity of the DE algorithm to find the global minimum.

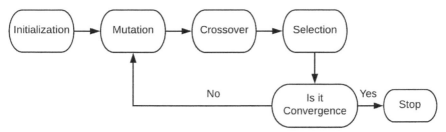

Figure 3.4 Flowchart of differential evolution algorithm (Adapted from [8]).

```
In[3]:=   NMinimize[{f[x1,x2],-35<x1<35,-35≤x2≤35},{x1,x2},
          Method->"DifferentialEvolution"]
Out[3]=   {2.2587*10^-9, {x1 -> -1.63413*10^-9,
          x2 -> -7.81672*10^-9}}
```

Changing *ScalingFactor* from defualt value of 0.6 to 0.7 obtained better results considering global optima.

```
In[5]:=   NMinimize[{f[x1,x2],-35≤x1≤35,-35≤x2≤35}, {x1,x2},
          Method->{"DifferentialEvolution",
          "ScalingFactor"-> 0.7}]
Out[5]=   {3.74914*10^-10, {x1 -> 8.68579*10^-10,
          x2 -> -1.00129*10^-9}}
```

Here, adjusting *ScalingFactor, RandomSead, CrossProbabilty* or *SearchPoints* didn't produce better global optima. Therefore they were kept at a default.

The global minima of other test function Holder Table 1 were tried to find by the algorithm. Initial steps are same with previously used algorithms.

```
In[10]:=  NMinimize[{f[x1,x2],-10≤x1≤10,-10≤x2≤10},{x1,x2},-
          Method->"DifferentialEvolution"]
Out[10]=  {-26.9203, {[x1 -> 9.64617, x2 -> -9.64617}}
```

In this example of function, all of the parameters being different than their defualt values did not obtain neither different results nor better global minima.

Nelder Mead

Nelder Mead algorithm (NM) or Simplex is one of the derivative-free optimization methods among other traditional local search algorithms. It was firstly designed for unconstrained optimization problems [8]. In m-dimensional space and given a function of m variables, this method keeps a set of m+1 points generating the vertices of a polytope. It should be noted that it must not be confused with the simplex method for linear programming. Iterations have performed by forming m+1 points as y_1, y_2, y_3,..., y_{m+1}. These points form the functions are ordered as $h(y_1) \leq h(y_2) \leq h(y_3) \leq ...h(y_{m+1})$. After the new point is produced to change with the previous worst point y_{m+1}. A polytope can be defined in terms of its centroid $(c = \Sigma_{i=1}^{m} y_i)$ being the average position of all the points of an object. Here, a trial point should be defined (y_t). It is produced by reflecting the worst point until centroid, $y_t = c + \alpha(c - y_{m+1})$ where α is a parameter being larger than 0. In this part, the new point need not be a new worst point or a new best point. Hence, $h(y_1) \leq h(y_t) \leq h(y_m)$, y_t replace with y_{m+1}. After obtaining a new point being better than the previous best point, it means that reflection is successfully obtained. Further, it can be continued with $y_e = c + \beta(y_t - r)$ where β being larger than 1 is a parameter to largen polytope. If $h(y_e)$ is obtained as smaller than $h(y_t)$, it means that the expansion process is achieved. Therefore, y_e changes with y_{m+1}. Alternatively, else, y_t changes as y_{m+1}. Another certain step for the algorithm process is that if the new point y_t underperforms to the second-worst point, $h(y_t) \leq h(y_m)$, the polytope is thought as very large and it is required to be contracted. Hence, a new trial point is obtained using the following expressions [9].

$$y_c = \begin{pmatrix} c + \gamma\,(y_{m+1} - c), & \text{if } h(y_t) \geq h(y_{m+1}) \\ c + \gamma(y_t - c), & \text{if } h(y_t) < h(y_{m+1}) \end{pmatrix}$$

where γ is a parameter ranging between 0 to 1. If contraction is achieved, it means that $h(y_c)$ is smaller than $\text{Min}[h(y_{m+1}), h(y_t)]$. Reversely, more process is required to obtain strong contraction.

Nelder-Mead has specific flexible options similar to other algorithms which are *ContractRatio, ExpandRatio, InitialPoints, PenaltyFunction, PostProcess, RandomSeed, ReflectRatio, ShrinkRatio,* and *Tolerance*. Even though this algorithm does not provide complete specifications that an accurate global optimization method should require, it tends to work well for the problem having less local minima. As previous algorithms, Nelder-Mead is used to obtain optimum global values for Ackley and Holder Table 1 test functions [1].

```
In[5]:=   NMinimize[{f[x1,x2],-35≤x1≤35,-35≤x2≤35},{x1,x2},
          Method->"NelderMead"]
Out[5]=   {0.87404, {x1 -> -0.998405, x2 -> -2.99522}}
```

It can be seen that results of the first trial are outperformed by DE while it gives better global optima compared to Random Search and Simulated Annealing solutions for Ackley function with the default set.

RandomSeed, which is referred to as one of the critical adjustment parameters of NM might directly affect the performance of the NM finding global minima.

```
In[6]:=  Do[Print[NMinimize[{f[x1,x2],-35≤x1≤35,-35≤x2≤35},
         {x1,x2},Method->{"NelderMead",
         "RandomSeed"->i}]],{i,5}]
Out[6]=  {0.557056,{x1->8.15872*10^-25,x2->-1.99487}}
         {0.280127,{x1->0.9948,x2->-6.32493*10^-9}}
         {7.12481,{x1->-20.9977,x2->-22.9975}}
         {2.32486*10^-10,{x1->4.63269*10^-10,
         x2->-6.78982*10^-10}}
         {1.3908,{x1->-4.99519,x2->-0.999038}}
```

Adjusting *RandomSeed* parameters provided a better minimum value of $2.32486 * 10^{-10}$, compared to trial performed with default set.

In this algorithm, other possible useful adjustment parameters are referred to as *ShrinkRatio, ContractRatio,* and *ReflectRatio*. However, it did not obtain global minima in Ackley function, as indicated the following.

```
In[7]:=  Do[Print[NMinimize[{f[x1,x2],-35≤x1≤35,-35≤x2≤35},
         {x1,x2},Method->{"NelderMead", "ShrinkRatio"->0.95,
         "ContractRatio"->0.95,"ReflectRatio"->2,
         "RandomSeed"->i}]],{i,5}]
Out[7]=  {0.39531,{x1->-0.996345,x2->-0.996345}}
         {0.783523,{x1->-1.99642,x2->1.99642}}
         {7.37952,{x1->-5.99939,x2->-31.9967}}
         {0.39531,{x1->-0.996345,x2->0.996345}}
         {2.40704*10^-9,{x1->-2.92841*10^-9,
         x2->-7.99045*10^-9}}
```

Another test function Holder Table 1 was minimized using the *NMinimize* command. As seen below, the global minima with default values were −26.9203.

```
In[12]:=  NMinimize[{f[x1,x2],-10≤x1≤10,-10≤x2≤10},{x1,x2},-
          Method->"NelderMead"]
Out[12]=  {-26.9203, {x1 -> 9.64617, x2 -> 9.64617}}
```

As applied for previous test function, firstly *RandomSeed* have been adjusted to find global minima.

```
In[13]:=   Do[Print[NMinimize[{f[x1,x2],-10≤x1≤10,-10≤x2≤10},
           {x1,x2},Method->{"NelderMead",
           "RandomSeed"->i}]],{i,5}]
Out[13]=   {-26.9203,{x1->-9.64617,x2->-9.64617}}
           {-9.13635,{x1->3.24199,x2->-9.71802}}
           {-26.9203,{x1->9.64617,x2->-9.64617}}
           {-7.76664,{x1->2.08542*10^-8,x2->9.73295}}
           {-7.76664,{x1->-7.64705*10^-9,x2->-9.73295}}
```

Outputs of this trial showed that adjustment of *RandomSeed* as is was not sufficient to reach minimum value. Lastly, other possibly useful parameters concerning literature for *NelderMead* were adjusted to obtain global minima.

```
In[14]:=   Do[Print[NMinimize[{f[x1,x2],-10≤x1≤10,-10≤x2≤10},
           {x1,x2},Method->{"NelderMead", "ShrinkRatio"->0.95,
           "ContractRatio"->0.95,"ReflectRatio"->2,
           "RandomSeed"->i}]],{i,5}]
Out[14]=   {-26.9203,{x1->-9.64617,x2->-9.64617}}
           {-26.9203,{x1->-9.64617,x2->-9.64617}}
           {-26.9203,{x1->-9.64617,x2->-9.64617}}
           {-26.9203,{x1->-9.64617,x2->-9.64617}}
           {-26.9203,{x1->-9.64617,x2->-9.64617}}
```

In this example, it was seen that none of the parameters could assure that a global minimum is different from the value obtained using the default settings.

References

[1] Wolfram Mathematica software 11.3.
[2] Ozturk, S., Aydin, L., Kucukdogan, N. and Celik, E. 2018. Optimization of lapping processes of silicon wafer for photovoltaic applications. Sol. Energy 164: 1–11.
[3] Momin, J., Xin-She, Y. and Hans-Jürgen, Z. 2013. Test functions for global optimization: A comprehensive survey. pp. 193–222. *In*: Xin-She, Y., Zhihua, C., Renbin, X., Amir Hossein, G. and Mehmet, K. [eds.]. Swarm Intelligence and Bio-Inspired Computation, Elsevier.
[4] Karaboğa, D. 2018. Yapay zeka optimizasyon algoritmaları. Nobel akademik yayıncılık, Ankara.
[5] Ingber, L. 1993. Simulated annealing: Practice versus theory. Math. Comput. Model. 18(11): 29–57.
[6] Savran, M. and Aydin, L. 2018. Stochastic optimization of graphite-flax/epoxy hybrid laminated composite for maximum fundamental frequency and minimum cost. Eng. Struct. 174: 675–687.

[7] Storn, R. and Price, K. 1997. Differential evolution—A simple and efficient heuristic for global optimization over continuous spaces. Journal of Global Optimization 11(4): 341–359. https://doi.org/10.1023/A:1008202821328.

[8] Vo-Duy, T., Ho-Huu, V., Do-Thi, T.D., Dang-Trung, H. and Nguyen-Thoi, T. 2017. A global numerical approach for lightweight design optimization of laminated composite plates subjected to frequency constraints. Composite Structures 159: 646–655.

[9] Nelder, J.A. and Mead, R. 1965. A simplex-method for function minimization. Computer Journal 7(4): 308–313. https://doi.org/10.1093/comjnl/7.4.308.

CHAPTER 4

Design and Optimization of Glass Reinforced Composite Driveshafts for Automotive Industry

Melih Savran,[1] *Ozan Ayakdaş,*[2] *Levent Aydin*[1],* and *M Eren Dizlek*[3]

Introduction

Fiber-reinforced composite materials are used in the many industrial applications such as aerospace, automotive, and marine due to them having a lightweight and giving superior performance. They give many design possibilities utilizing the commutative stacking sequences and fiber orientation angles. Since the effects of the design parameters and stacking sequences are very complicated with the possible interactive relations among them, stochastic optimization approaches take a more active role for the composite design problems. On the other hand, usage of the composites had been started being employed instead of the leading traditional materials in the industry since the few last decades. Especially, since fiber-reinforced composite materials have gained significant importance in the automotive industry because of their relatively superior physical properties and main characteristics of them [1]. Fiber-reinforced composites generally composed of glass, carbon or hybrid (carbon &

---
[1] İzmir Katip Çelebi University, Mechanical Engineering Department, İzmir, Turkey.
  Email: mlhsvrn@gmail.com
[2] İzmir Institute of Technology, Mechanical Engineering Department, İzmir, Turkey.
  Email: ozanayakdas@gmail.com
[3] R&D Manager/Design Center Director at Manisa Kardan CEMMER A.Ş., Manisa.
  Email: erendizlek@hotmail.com
* Corresponding author: leventaydinn@gmail.com

glass) fiber in the automotive main parts are due to having high strength in the fiber direction. These materials are durable and rigid; moreover, that might be saved in cost and weight thanks to optimum design.

In the last decade with the increase in the obligation of this arrangement on the environment, the automotive industry is faced with many severe sanctions such as reduction of weight and energy consumption, low CO_2 emissions without compromise in vehicle quality. All of the problems related to weight reduction and energy consumption can be overcome by using composite materials, the addition of this it is a fact that the weight of a vehicle and its fuel consumption are proportional [2].

In a vehicle, the transfer of the torque produced by the motors to the wheels is generally solved by using a "power transmission system." This system may be changed according to the driving type of vehicle [3]. In the power transmission system, rigid driveshafts are the main parts which are transferred to the torque from the gearbox to the axle/differential (see Figure 4.1).

Driveshafts are generally made up from steel or aluminum tubes subjected to torsion or shear stress and have been used in the automobiles or other vehicles together with fundamental design rules. These main

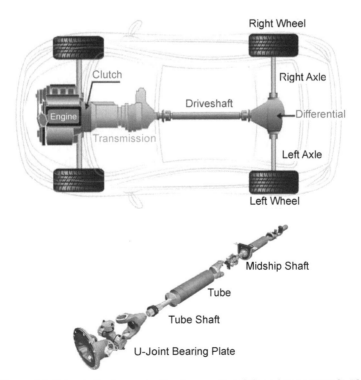

Figure 4.1 Driveshafts demonstration in an automobile and its sections [4, 5].

design rules are said to have enough torsional strength and to have a natural bending frequency. In addition to these rules, it should carry torque without failure, and the thin-walled tube geometry should carry torque without failure and buckling [6]. Engineers generally produce driveshafts in two pieces from steel or aluminum to provide a relatively high bending natural frequency. Also, to eliminate the disadvantage of the low bending natural frequency of long shafts with excessive tube lengths, drive shafts are generally spitted in two or more sections and supported with center bearings (with a rubber damper) between sections. Nowadays, due to being critical towards the reduction in inertial mass, manufacturing of the composite power transmission shaft is an attractive issue. Many researchers are interested in composite materials for automotive drive shaft applications in place of steel or aluminum materials because composite shafts can be more efficient with regards to the shock absorber and therefore decrease the stress created. For instance, an optimization problem including a comparison of the steel and composite driveshaft has been carried out by [7]. It is shown that to use carbon fiber reinforced composites instead of traditional structural steel reduced weight and stresses in the drive shaft. In another optimization study of the composite driveshaft, the effect of the stacking sequences of composite fiber orientation angles has been found to be very active on buckling torque [8].

Additionally, a significant reduction has been seen in weight for using composite rather than steel driveshaft. In a weight comparison study on the lightweight composite and steel driveshaft, carbon/epoxy and kevlar/epoxy composite driveshaft provide 89.756% and 72.53% of weight reduction, respectively, instead of SM45C steel driveshaft [9]. In the view of the buckling strength for composite driveshafts, it is observed that carbon/epoxy composite driveshaft has excellent performance [10]. As a brief of the literature, composite driveshafts show enough performance, when various loading types such as shear strength and bending natural frequency are considered. The impact resistance of the composite driveshaft has also been investigated in the study by Henrya [11]. Additionally, this optimization design logic has been implemented on the development part of the study. In the study related to the effects of carbon/epoxy winding layer on torsional characteristics of filament wound composite shafts, it has resulted that in the comparison of the helix and hoop layer, helix winding layer provides high hardness and more resistance to torsional forces [12].

Among the fiber-reinforced composite materials studied in the literature, graphite fiber reinforced ones are mostly known due to having superior properties such as (i) high specific strength and modulus, (ii) high fatigue strength, (iii) low coefficient of thermal expansion. However, due to carbon fibers proving to be expensive, glass fibers are used as an alternative in automobile parts, and they have a lower cost, higher strength, low elastic modulus, low fatigue strength, and high density. Although glass fibers are not as high strength as carbon fibers, safe designs can be satisfied by using stochastic optimization methods. There are many kinds of research on composite winding thin aluminum cylinders driveshafts. These driveshafts are composed of the aluminum inner liner and composite winding parts. It is seen that composite wound driveshafts demonstrate better strength and vibration performance at lower weight compared to steel driveshafts. For instance, a one-piece composite drive shaft consisting of graphite/epoxy and aluminum-graphite/epoxy has been investigated by replacing a two-piece steel shaft [13]. Similarly, in another study on the design and development of the aluminum-glass fiber driveshaft, it was shown that increasing the number of layers would improve the maximum static torsion approximately 66% for [+45/−45] laminates [14]. In addition to this result, it can be said that the number of composite layers and their fatigue strength can be proportional for the aluminum/composite driveshaft.

High natural frequency and a high buckling load resistance have been considered in a study which compares conventional steel and composite driveshafts [15]. As a result of the study, to use composite material in driveshaft has provided a decline in weight by 50% while increasing frequency by 14% at the same loading capacity. In another study [16], the composite automotive driveshaft problem composed of hybrid carbon and glass fiber/epoxy have been investigated. The following results were proposed: (i) failure and fatigue capability of the driveshaft can be changed by stacking sequence design, (ii) to increase winding angle does not affect the buckling resistance systematically. At this point, the importance of optimization can appear clearly in the design of composite driveshaft.

There are many complicated-highly nonlinear functions in the design and optimization problems of the composite driveshaft. Therefore, stochastic optimization methods such as Differential Evolution, Genetic Algorithm, and Simulated Annealing are superior towards solving these problems. Researchers have recently focused on problems related to using stochastic algorithms on the design of the driveshaft problems. For example, in the study by Dinesh and Raju [17], the Genetic Algorithm is run to obtain the appropriate stacking sequence with the weight minimization problem for glass/epoxy and carbon/epoxy composite driveshaft. In another study, the fundamental frequency and critical buckling load have been maximized while minimizing the weight using Modified Non-dominated Sorting Genetic Algorithm (NSGA II) for hybrid graphite-glass/epoxy composite shaft [15]. Additionally, many researchers have also utilized the finite element method to propose safety designs for a composite drive shaft with different mechanical parameter calculations [1, 8, 17, 15].

In the present study, in order to minimize weight, stacking sequence optimization of composite driveshafts are considered using various glass/epoxy materials employing the Differential Evolution method (see Chapter 2 for detailed information). Bending natural frequency, torsional buckling, and manufacturing constraints are taken into account for the design. Fiber orientations, the numbers, and the thicknesses of the layers are selected as design variables.

Design Procedure

The mechanical behavior of the composite driveshaft is determined by using Classical Laminated Plate theory. The theory considers the following assumptions:

1. Each lamina is elastic and orthotropic.
2. Each lamina is homogeneous and perfectly bounded each other.
3. The laminated composite is thin, and the thickness of the composite plate is much lesser than its edge dimensions.
4. The laminated composite (except for their edges) is considered as in-plane stress ($\sigma_z = \tau_{xz} = \tau_{yz} = 0$). The loadings are only applied in the laminate's plane.
5. As compared to the thickness of the laminate, displacements are tiny and continuous throughout the thickness.
6. In the x and y directions, in-plane displacements are linear functions of z.
7. Transverse shear strains (γ_{xz} and γ_{yz}) are ignorable for the reason that the line is straight and perpendicular to the middle surface.

First of all, the elements of the transformed reduced stiffness matrix $[\bar{Q}_{ij}]$ can be written as in the following form by using the mechanical properties of composite materials for an angle lamina [18].

$$\bar{Q}_{11} = Q_{11}c^4 + Q_{22}s^4 + 2(Q_{12} + 2Q_{66})s^2c^2$$

$$\bar{Q}_{12} = (Q_{11} + Q_{22} - 4Q_{66})s^2c^2 + Q_{12}(c^4 + s^4)$$

$$\bar{Q}_{22} = Q_{11}s^4 + Q_{22}c^4 + 2(Q_{12} + 2Q_{66})s^2c^2$$

$$\bar{Q}_{16} = (Q_{11} - Q_{12} - 2Q_{66})sc^3 - (Q_{22} - Q_{12} - 2Q_{66})s^3c$$

$$\bar{Q}_{26} = (Q_{11} - Q_{12} - 2Q_{66})cs^3 - (Q_{22} - Q_{12} - 2Q_{66})sc^3$$

$$\bar{Q}_{66} = (Q_{11} + Q_{22} - 2Q_{12} - 2Q_{66})s^2c^2 + Q_{66}(c^4 + s^4) \qquad (4.1\text{–}4.6)$$

where stiffness matrix quantities $[Q_{ij}]$ including engineering constants E_1, E_2, v_{12} and G_{12} are

$$Q_{11} = \frac{E_1}{1 - v_{21}v_{12}} \quad Q_{22} = \frac{E_2}{1 - v_{21}v_{12}} \quad Q_{66} = G_{12} \tag{4.7}$$

$$Q_{12} = \frac{v_{12}E_2}{1 - v_{21}v_{12}} \quad v_{21} = \frac{E_2}{E_1}v_{12}$$

Secondly, by using the above equations, extensional stiffness matrix [A] for the whole composite plate can be determined as [18]

$$A_{ij} = \sum_{k=1}^{n} [(\overline{Q}_{ij})]_k (h_k - h_{k-1}), \; i, j = 1, 2, 6 \tag{4.8}$$

where h_k is the locations of the ply surfaces, n is the number of layers.

E_x and E_h in the axial and hoop directions can be defined by combined elements of the matrix $[A_{ij}]$ and the total thickness and given respectively as [19].

$$E_x = \frac{1}{t}\left[A_{11} - \frac{A_{12}^2}{A_{22}} \right] \qquad E_h = \frac{1}{t}\left[A_{22} - \frac{A_{12}^2}{A_{11}} \right] \tag{4.9}$$

where A_{11}, A_{12}, and A_{22} are the corresponding elements of matrix A_{ij}, t represents the total thickness of the layers.

Buckling Torque

An orthotropic thin-walled hollow cylindrical shaft exhibits torsional buckling behavior just after the applied torque exceeds the critical torsional buckling load T_{cr} [20]:

$$T_{cr} = (2\pi r_m^2 t)(0.272)(E_x E_h^3)^{1/4}\left(\frac{t}{r_m} \right)^{3/2} \tag{4.10}$$

where r_m, is the mean radius of the cylinder.

It is seen that torsional buckling is proportional to $E_x^{1/4}$ and $E_h^{3/4}$. It can be seen that 90-degree layers should be added to increase the torsional buckling of the composite driveshaft.

Lateral Bending Natural Frequency

Natural frequency is the most crucial issue which effects on the driveshafts for the dynamical engineering system. Therefore, to avoid any adverse effects on vehicle NVH (Noise, Vibration, Harshness), the natural frequency must be higher than the frequency of the vibration sources on the vehicle. Internal combustion engine is the main vibration source in the vehicle so, designers have to design drive shafts which have a higher natural frequency (f_n) than engine vibration range. It can be expressed as in the following form [20]:

$$f_n = \frac{\pi}{2}\sqrt{\frac{g\,E_x\,I}{W_u\,L^4}} \tag{4.11}$$

where g is the gravity acceleration, W_u is the weight per unit length; I is the moment of inertia, L is the length of the shaft. For the thin-walled tube, the moment of inertia can be expressed as

$$I = \frac{\pi}{4}\left(r_0^4 - r_i^4\right) \tag{4.12}$$

where r_0 and r_i is the outer and inner radius of the drive shaft, respectively.

Design Problem and Formulation

The considered composite driveshaft is an element of an automobile power transmission system and represents an orthotropic thin-walled hollow cylindrical with 50 mm internal radius and length of 1480 mm (see Figure 4.2).

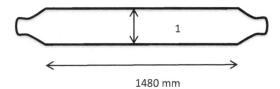

1

1480 mm

Figure 4.2 Fiber composite drive shaft [20].

The design problems of the composite drive shaft have been formulated as follows:

Minimize:
Weight$=\pi(r_0^2 - r_i^2)\rho$

Such that:
$f_n \geq 80$ Hz (frequency); $T_{cr} \geq 550$ Nm (buckling);
$\Theta_i \in \{0°, \pm 45°, 90°\}$ (fabrication)

Design variables
$\Theta_1, \Theta_2, ..., \Theta_m, n, t$ and material type

where ρ is the mass density, $\Theta_1, \Theta_2, ..., \Theta_m$ are fiber orientation angle of glass/epoxy layers.

In this section, composite driveshaft design has been conducted under frequency, buckling, and manufacturing constraints. Firstly, in order to validate the proposed algorithm MDE (Modified Differential

Evolution), the problem given in the Kaw [20] is solved for glass/epoxy composite drive shaft, and then, as an original optimization problem, weight minimization of composite drive shafts consisting of various glass-epoxy materials given in the literature are considered. Thus, the effect of mechanical properties of various glass/epoxy materials given in literature on composite driveshaft design has been observed. The mechanical properties of considered driveshaft materials are given in Table 4.1.

Table 4.1 Properties of different glass/epoxy materials.

Material No.	Reference Studies	E_1 (Gpa)	E_2 (Gpa)	G_{12} (Gpa)	v_{12}	ρ (kg/m³)
1	[21]	37	8.5	4.7	0.254	1800
2	[22]	21.97	4.13	0.518	0.31	2100
3	[23]	30.90	8.30	2.80	0.32	1800
4	[24]	40	10	3.15	0.3	1780
5	[25]	46.5	12.1	4.47	0.36	1760
6	[26]	50	12	5.6	0.3	2000
7	[10]	40	8	4	0.25	1900
8	[27]	39	8.6	3.8	0.28	2100
9	[28]	44	10.5	3.74	0.36	1506
10	[29]	40.51	13.96	3.10	0.22	1830
11	[30]	18.3	7.94	3.895	0.25	1550
12	[31]	40.3	6.21	3.07	0.2	1910
13	[32]	39	8.6	2.5	0.28	2100
14	[33]	53.78	17.93	8.96	0.25	1900
15	[34]	31	12	3.20	0.3	1600
16	[35]	45.6	16.2	5.83	0.278	2080
17	[36]	41.25	9.24	3.38	0.26	1270
18	[37]	38.6	8.27	4.14	0.26	1800

Optimization

Optimization techniques can be categorized into two headings as traditional and non-traditional optimization methods. Traditional optimization techniques like Constrained Variation and Lagrange Multipliers, which are analytical, can achieve the optimum solution of differentiable and continuous functions. Furthermore, traditional optimization techniques cannot be utilized to solve composite design

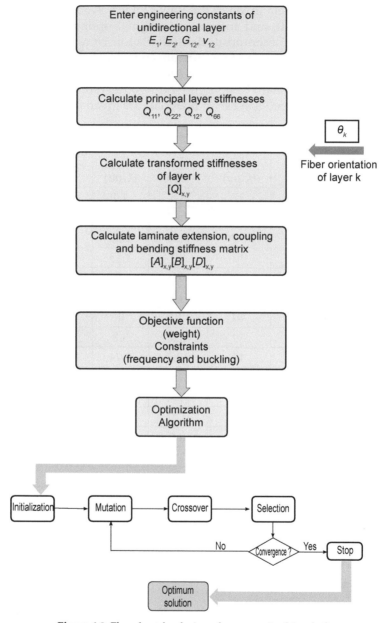

Figure 4.3 Flowchart for design of a composite driveshaft.

problems due to containing discrete search spaces and nonlinear functions [38]. For these reasons, stochastic optimization methods such as Genetic Algorithms (GA), Differential Evolution (DE), Nelder-Mead (NM) and

Simulated Annealing (SA) can be used effectively [39]. A review of the various optimization methods and algorithms have been summarized in [40] and [41] for engineering applications and the problems of composite design. Among these algorithms, the DE method is selected to solve the optimizations problems of composite drive shafts. It is noticed that the Modified Differential Evolution algorithm (MDE) which is different from the standard DE algorithm used in this study. In the analysis based on the Mathematica software, "PostProcess" and "PenaltyFunction" add-ons provide for the algorithm to be more reliable and more potent than traditional DE in finding the global optimum. The way to solve the optimization problem is summarized in the flow chart by including MDE.

Numerical Results

In this part, the results of the optimization problems are given based on the MDE (Modified Differential Evolution) method. In order to validate the optimization code used in this study, the problem which is given in [1] is solved by using the MDE method and then compared. The results for the stacking sequence designs of symmetric glass/epoxy composite driveshafts for minimum weight using the proposed MDE algorithm and PROMAL program by Kaw [20] are given in Table 4.2. The corresponding parameters "number of plies," "critical torsional buckling load," "stiffnesses," and "bending natural frequency" are also calculated based on optimal laminate stacking sequence designs. It can be seen that MDE and PROMAL give the same results for the problems and hence, MDE shows potential to obtain reliable results for similar design problems.

Table 4.3 shows the effect of different glass-epoxy materials on lightweight drive shaft design. The utilizing of seventeen glass-epoxy composites studied in literature are selected as design materials. The weigh values of glass-epoxy composite drive shafts vary in the range of 0.7289–2.3800 kg under the constraints, critical torsional buckling load $T_{cr} > 550$ Nm and bending natural frequency $f_n > 80$ Hz. According to

Table 4.2 Comparison between the results of [20] and the present study for glass/epoxy composite driveshaft with the design constraints $f_n \geq 80$ Hz, $T_{cr} \geq 550$ Nm.

Method	Laminate Stacking Sequence	Number of Plies	T_{cr} (Nm)	E_x (GPa)	E_y (GPa)	f_n (Hz)	Weight (kg)
PROMAL [1]	$[0_2/\pm45_2/90]_s$	14	828.8	20.16	16.16	83.7	1.405
	$[45/90_3/0_2]_s$	12	763.5	9.44	24.47	82.4	1.226
MDE (Present Study)	$[0_2/\pm45_2/90]_s$	14	828.8	20.16	16.16	84.1	1.413
	$[45/90_3/0_2]_s$	12	763.6	19.44	24.47	82.7	1.218
	$[45/0/90/0/90/\underline{0}]_s$	11	560.6	23.16	20.42	89.7	1.135

Table 4.3 Stacking sequence designs of the driveshaft for minimum weight using different glass/epoxy materials.

Material No.	Laminate Stacking Sequence	No. Plies	Critical Torsional Buckling Load (Nm)	Minimum Natural Frequency (Hz)	Weight (kg)
1	$[90/0_3/90/\underline{90}]_s$	11	590,7790	91.6711	1.1349
2	$[0_{10}]_s$	20	739,6160	79.9842	2.3800
3	$[90/0/90_2/0/90_2/0_3/\underline{90}]_s$	21	2669,2800	80.7481	2.1393
4	$[90/0/90/0/90]_s$	10	557,4970	88.3833	1.0216
5	$[0_3/90_2]_s$	10	582,2010	108.6260	1.0101
6	$[0/90_3/\underline{90}]_s$	9	565,0430	80.4160	1.0344
7	$[90_2/0/90/0/\underline{0}]_s$	11	660,6900	86.3865	1.1980
8	$[0_2/90_3/\underline{0}]_s$	11	655,4500	82.0004	1.3241
9	$[90_3/0_2]_s$	10	610,5800	100.3340	0.8643
10	$[0_2/90_3]_s$	10	599,4600	91.9465	1.0503
11	$[0_6/90/\underline{90}]_s$	15	652,9220	80.6336	1.3259
12	$[0/90_2/0/90/\underline{0}]_s$	11	640,4180	84.4438	1.2043
13	$[90/0_2/90_2/\underline{0}]_s$	11	655,4500	82.0004	1.3241
14	$[0/90/0/90/\underline{0}]_s$	9	564,6050	112.1870	0.9827
15	$[0/90/0/90/0/\underline{90}]_s$	11	564,1240	93.8224	1.0088
16	$[0/90_3/\underline{90}]_s$	9	553,0140	83.0797	1.0758
17	$[90/0/90_2/0]_s$	10	563,2830	104.6340	0.7289
	Design constraints		> 550	> 80	

results in Table 4.3, Material 17 has a minimum weight (0.7289 kg) while comparing with others while with the usage of Material 2, glass-epoxy has obtained the heaviest drive shaft (2.3800 kg). By using Material 17 instead of Material 2, drive shaft weight can be decreased up to 69%.

A close look at the results given in Table 4.3 demonstrates that although a 45-degree fiber angle increment is selected as a design constraint, the obtained stacking sequence configuration of laminated-glass epoxy composite driveshaft only includes 0 and 90-degree angles.

(i) By considering the design constraints $T_{cr} > 550$ Nm, $f_n > 80$ Hz, and Equation (4.10), torsional buckling proportional with $E_h^{3/4}$ and $E_x^{1/4}$. In here, elastic modulus in h-direction "E_h" is more effective in the increment of torsional buckling, so 90-degree fiber angle is to be dominant, (ii) Similarly, according to Equation (4.12), bending natural frequency proportional with E_x and 0-degree fiber angle is to be found as an appropriate value by the MDE optimization method. Effects of fiber orientation angles on f_n, T_{cr}, E_h, and E_x relation with fiber angles are given in Figure 4.4.

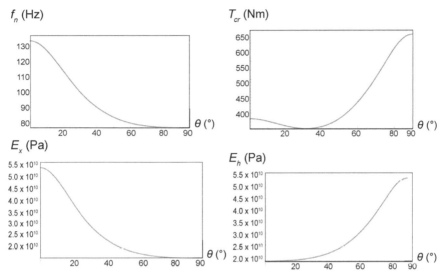

Figure 4.4 Bending natural frequency "f_n," torsional buckling "T_{cr}," x and h directions modulus "E_x and E_h" as a function of angle of laminated composite.

When the results are examined, it seems that the highest critical torsional buckling value (2669.28 Nm) is obtained using material 3, while material 16 provides the lowest ones. Also, the highest and lowest bending natural frequency (112.1870 Hz, 80.4160 Hz) acquire utilizing material 14 and 6, respectively. When all the obtained results are compared, it can be said that the selection of which glass-epoxy materials will be used in the drive shaft design is a vital process towards reducing weight.

Conclusion

In this chapter, a comparison of drive shaft designs for various glass-epoxy composite materials given in literature has been conducted using bending natural frequency, torsional buckling, and manufacturing constraints. Minimum weight of composite driveshafts is considered. The use of different glass-epoxy composite materials ensures a considerable amount of weight reduction for drive shafts.

It can be concluded that

1) When compared to glass/epoxy materials given in literature in terms of lightweight driveshaft design, the drive shaft weight changes in the range of 0.7289 kg–2.3800 kg.
2) 0 and 90-degree fiber angles are more effective for drive shaft design.

3) By using different glass-epoxy materials, drive shaft weight can be decreased up to 69%.
4) The selection of which glass-epoxy materials will be used in the drive shaft design is a critical process in order to reduce weight.

References

[1] Khoshravan, M.R. and Paykani, A. 2012. Design of a composite drive shaft and its coupling for automotive application. Journal of Applied Research and Technology 10: 826–834.

[2] Dhanwate, S., Pimpale, S. and Kulkarni, S. 2008. Design & optimization of automotive composite drive shaft. International Journal of Scientific Research and Management Studies 1: 413–421.

[3] Heisler, H. 1999. Vehicle and Engine Technology, SAE International, London.

[4] Dana commercial vehicle products, http://www.dana.com/commercial vehicle/ products/driveline/driveshafts.

[5] Kannan, V., Kannan, V.V. and Pemmasani, S. 2018. Design optimization of an epoxy carbon prepreg drive shaft and design of a hybrid aluminium 6061-t6 alloy/epoxy carbon prepreg drive shaft. SAE Technical Paper-28-0014.

[6] Mazziotti, P.J. 1954. Universal joint and propeller shaft. Dana Corporation Bulletin, J-1371.

[7] Narayana, V.I., Mojeswararao, D. and Kumar, M.N.V.R.L.K. 2012. Material optimization of composite drive shaft assembly in comparison with conventional steel drive shaft. International Journal of Engineering Research & Technology 1: 1–9.

[8] Rangaswamy, T., Vijayarangan, S., Chandrashekar, R.A., Venkatesh, T.K. and Anantharaman, K. 2004. Optimal design and analysis of automotive composite drive shaft. International Symposium of Research Students on Materials Science and Engineering, 1–9.

[9] Moorthy, R.S., Mitiku, Y. and Sridhar, K. 2013. Design of automobile driveshaft using carbon/epoxy and kevlar/epoxy composites. American Journal of Engineering Research 2: 173–179.

[10] Madhu, K.S., Darshan, B.H. and Manjunath, K. 2013. Buckling analysis of composite drive shaft for automotive applications. Journal of Innovative Research and Solutions 1: 63–70.

[11] Henrya, T.C. and Mills, B.T. 2019. Optimized design for projectile impact survivability of a carbon fiber composite drive shaft. Composite Structures 207: 438–445.

[12] Tariq, M., Nisar, S., Shah, A. et al. 2018. Effect of carbon fiber winding layer on torsional characteristics of filament wound composite shafts. Journal of the Brazilian Society of Mechanical Sciences and Engineering 40: 198.

[13] Chowdhuri, M.A.K. and Hossain, R.A. 2010. Design analysis of an automotive composite drive shaft. International Journal of Scientific Research and Management Studies 2: 45–48.

[14] Arun, M. and Vinoth, K.S. 2013. Design and development of laminated aluminum glass fiber drive shaft for light duty vehicles. International Journal of Innovative Technology and Exploring Engineering 2: 157–165.

[15] Khalkhali, A., Nikghalb, E. and Norouzian, M. 2015. Multi-objective optimization of hybrid carbon/glass fiber reinforced epoxy composite automotive drive shaft. International Journal of Engineering 28: 583–592.

[16] Abu Talib, A.R., Ali, A., Badie, M.A., Che Lah, N.A. and Golestaneh, A.F. 2010. Developing a hybrid, carbon/glass fiber-reinforced, epoxy composite automotive drive shaft. Materials and Design 31: 514–521.

[17] Dinesh, D. and Raju, F.A. 2012. Optimum design and analysis of a composite driveshaft for an automobile by using genetic algorithm and Ansys. International Journal of Engineering Research and Applications 2: 1874–1880.

[18] Seydibeyoğlu, M.Ö., Mohanty, A.K. and Misra, M. 2017. Fiber technology for fiber-reinforced composites. pp. 5–22. *In*: Aydin, L., Artem, H.S., Oterkus, E., Gundogdu, O. and Akbulut, H. [eds.]. Mechanics of Fiber Composites. WoodHead Publishing, Elsevier, United Kingdom.

[19] Badie, M.A., Mahdi, A., Abu Talib, A.R., Abdullah, E.J. and Yonus, R. 2006. Automotive composite drive shafts: investigation of the design variables effects. International Journal of Engineering and Technology 3: 227–237.

[20] Kaw, A. 2006. Mechanics of composite materials. pp. 411–419. *In*: Kaw, A. [ed.]. Failure, Analysis, and Design of Laminates. CRC Press, Florida, USA.

[21] Asaee, Z. and Taheri, F. 2015. Experimental and numerical investigation into the influence of stacking sequence on the low-velocity impact response of new 3D FMLs. Composite Structures 140: 136–146.

[22] Zhang, Z. and Taheri, F. 2004. Dynamic pulse-buckling behavior of "quasi-ductile" carbon/epoxy and E-glass/epoxy laminated composite beams. Composite Structures 64(3-4): 269–274.

[23] Kathiresan, M., Manisekar, K. and Manikandan, V. 2014. Crashworthiness analysis of glass fibre/epoxy laminated thin walled composite conical frusta under axial compression. Composite Structures 108(1): 584–599. https://doi.org/10.1016/j.compstruct.2013.09.060.

[24] Singh, H., Namala, K.K. and Mahajan, P. 2015. A damage evolution study of E-glass/epoxy composite under low velocity impact. Composites Part B: Engineering 76: 235–248.

[25] Kalhor, R., Akbarshahi, H. and Case, S.W. 2016. Numerical modeling of the effects of FRP thickness and stacking sequence on energy absorption of metal-FRP square tubes. Composite Structures 147: 231–246.

[26] Dhanwate, S., Pimpale, S. and Kulkarni, S. 2012. Design & optimization of automotive composite drive shaft. International Journal of Scientific Research and Management Studies 1(12): 413–421.

[27] Isaac, M. and Daniel, O.I. 1994. Engineering Mechanics of Composite Materials. Oxford University Press.

[28] Okutan, B. and Karakuzu, R. 2002. The failure strength for pin-loaded multi-directional fiber-glass reinforced epoxy laminate. Journal of Composite Materials 36(24): 2695–2712.

[29] Icten, B.M., Atas, C., Aktas, M. and Karakuzu, R. 2009. Low temperature effect on impact response of quasi-isotropic glass/epoxy laminated plates. Composite Structures 91(3): 318–323.

[30] Ekşi, S. and Genel, K. 2017. Comparison of mechanical properties of unidirectional and woven carbon, glass and aramid fiber reinforced epoxy composites. Acta Physica Polonica A 132(3): 879–882.

[31] Badie, M.A., Mahdi, E. and Hamouda, A.M.S. 2011. An investigation into hybrid carbon/glass fiber reinforced epoxy composite automotive drive shaft. Materials and Design 32(3): 1485–1500.

[32] Madukauwa-David, I.D. and Drissi-Habti, M. 2016. Numerical simulation of the mechanical behavior of a large smart composite platform under static loads. Composites Part B: Engineering 88: 19–25.

[33] Tornabene, F., Fantuzzi, N. and Bacciocchi, M. 2016. On the mechanics of laminated doubly-curved shells subjected to point and line loads. International Journal of Engineering Science 109: 115–164.

[34] Okutan, B. 2017. Curved sandwich composites with layer-wise graded cores under impact loads. Composite Structures 159: 1–11. https://doi.org/10.1016/j.compstruct.2016.09.054.

[35] Gillet, A., Francescato, P. and Saffre, P. 2010. Single and multi-objective optimization of composite structures: The influence of design variables. Journal of Composite Materials 44(4): 457–480.

[36] Şahin, M. 2008. Burkulmaya maruz tabakali kompozit plaklarin deneysel ve nümerik analizi. M.S. Thesis, Cumhuriyet University, Sivas.

[37] Tsai, S. and Hahn, H. 1980. Introduction to composite materials, Wesport Technomic Publishing Company.

[38] Aydin, L. and Artem, H.S. 2011. Comparison of stochastic search optimization algorithms for the laminated composites under mechanical and hygrothermal loadings. Journal of Reinforced Plastics and Composites 30: 1197–1212.

[39] Seydibeyoğlu, M.Ö., Mohanty, A.K. and Misra, M. 2017. Fiber technology for fiber-reinforced composites. pp. 299–314. *In*: Aydin, L. and Artem, H.S. [eds.]. Design and Optimization of Fiber Composites. WoodHead Publishing, Elsevier, United Kingdom.

[40] Rao, S.S. 2009. Engineering Optimization: Theory and Practice. Wiley.

[41] Gurdal, Z., Haftka, R.T. and Hajela P. 1999. Design and Optimization of Laminated Composite Materials. John Wiley & Sons.

CHAPTER 5

Dual Mass Flywheel Optimization

Ümmühan Gümüş,[1] Levent Aydin,[2,]*
Okan Yazici[3] and Samet Kabacaoğlu[4]

Introduction

One of the crucial factors affecting vehicle performance is the torsional vibrations that occur in the clutch system and these vibrations also cause noise to occur in the power transmission. In today's technology, vehicle designers use dampers, couplings and similar products in order to reduce torsional vibrations or provide isolation [1]. Consequently, it becomes apparent that conventional damping elements do not provide sufficient vehicle comfort due to the limitations in damping and limited volumes in the installation. For these reasons, Dual Mass Flywheel appeared in the 1980s. In this respect, DMF is a highly efficient vibration damping element, widely used in passenger and commercial vehicles, more preferred in diesel vehicles [1]. The resonance value for torsion in traditional single mass flywheels is between 700 and 2,000 rpm. DMF aims to reduce the torsional resonance of the clutch system below the engine resonance. This is provided by having an additional mass on the transmission input shaft

[1] İzmir Katip Çelebi University, Graduate School of Natural and Applied Sciences, İzmir, Turkey.
  Email: ummuhang93@gmail.com
[2] İzmir Katip Çelebi University, Department of Mechanical Engineering, İzmir, Turkey.
[3] BMC (BMC Otomotiv Sanayi ve Ticaret A.Ş.), İzmir, Turkey.
  Email: uokanyazici@gmail.com
[4] Dönmez Debriyaj (Hammer Clutch UK, Hammer gmbh), İzmir, Turkey.
  Email: sametkabacaoglu@gmail.com
* Corresponding author: leventaydinn@gmail.com

side. DMF also aims to reduce the hysteresis in the system and to dampen the power fluctuations from the motor shaft [2]. The inclusion of DMF also increases driver comfort and allows the engine to run at lower speeds, resulting in lower fuel consumption [3].

Studies on the related area in the literature can be examined in three main sections.

Theoretical and Experimental

Subsequent studies have been based on various methods to investigate the effect of DMF on the transmission system, and the mathematical models have been established. In order to see the accuracy, the results of the experimental setups were compared. As the torsional stiffness changes depending on the rotational speed, it also affects the damping characteristic of DMF. Therefore, the effect of torsional stiffness is one of the topics studied [3–6].

There is a study that aims at further improving the counter-torque of a DMF with single-stage stiffness at large torsional angles. Besides, it is used to model the contact between the friction bearing and the secondary flywheel [3]. In a different study, it was found that the two stages of torsional stiffness were calculated considering the frictional forces. For Circumferential Arc Spring Dual Mass Flywheel (DMF-CS), the repetitive formulas of the springs in the system were calculated by a different method, including friction forces [4]. There are also studies on multistage DMF design methods on torsional vibration control and the compatibility of kinetic parameters. Studies have shown that the amplitude of the angular velocity in the gearbox input shaft is reduced by using three-stage DMF and torsional vibrations can be effectively controlled [5].

The friction characteristic of the arc springs is another important point to be emphasized. For this reason, studies have been carried out on this subject [4, 7]. The study by Kim et al. 2006, it is performed a discrete analysis approach for DMF performance and efficiency. The arc spring is modelled as an n-discrete element and is placed between the flywheels, and each element contains components such as mass, spring, and non-linear friction elements. In order to explain the viscous friction and Stribeck effect due to relative sliding velocity, nonlinear element model is defined. To define the arc spring friction behavior, they have introduced LuGre's model. MATLAB Simulink is also often used in such studies [7–9].

Alternatively, a new DMF structure with variable stiffness has been investigated in order to determine the characteristics of the power transmission system in [6]. The resonance speeds of the torsional stiffness in orders (first and second) have been analyzed, and the model is transformed into fourth-order linear algebraic equations. Since DMF cannot be defined as a linear dynamic model, a suitable model should be selected to determine the ignored non-linear properties, which is the creation of a non-linear dynamic model [10, 11]. The researchers consider the non-linear stiffness, and the parameters in the Bouc–Wen model (the hysteresis caused by friction torque) are estimated by obtaining dynamic test data. When the Levenberg-Marquardt and Gauss-Newton method were compared, the first method showed more efficient and accurate results [10]. In the Coulomb method, a quasi-steady state approach was used to simplify the complexity of the model for arc springs [11]. In the DMF model, the observability of the motor torque was studied by using the Ab-initio method [12]. In another study, optimization of DMF and clutch system was emphasized. It was found that the theoretical studies were compatible with the experimental models, and vibration was reduced by the optimization of the parameters required for the design [10]. Many methods have been used in this subject, and the theoretical studies have agreed well with experimental studies and contributed to its development. Therefore, the contribution of the studies carried out in this field is significant.

Experimental

By establishing test benches and models, the effects of torsional vibrations on the transmission of a dual-mass flywheel (DMF) and noise and vibration improvements were investigated [2, 13–15]. In a study involving many of the powertrain components, torsional vibration signals were obtained by equal angles sampling and order analysis method. The torsional vibration data are processed and analyzed with the help of MATLAB® program under constant speed conditions, and the order signal of the engine excitation is obtained [13]. In another study, a test setup was established to investigate the effect of DMF on impact-induced clonk noise. Extensive measurements using DMF and the conventional clutch system showed significant differences in vibration and noise values in the driveline, resulting in differences in the number of modes of both parameters [2]. Dual Clutch Transmission (DCT) DQ250 was used to observe and verify the damping ability of the DMF, and the experimental setup was established. In this experiment, various conditions were created and tested for the vehicle, which can be given as an idle, working and climbing condition. According to the obtained test data, different engine revs and loads affect the damping ability of the DMF. As the load and rev on the engine increases, the dual mass flywheel has been observed to have a better damping ability [14]. Displacement measurements on DMF have been developed with a linear variable differential transformer (LVDT) mounted on a vehicle. Unlike other experimental studies, the data are obtained by using sensors, power and Bluetooth TM node connection and transmits the data via Bluetooth technology. It is possible to obtain more direct results with this measurement system developed for DMF.

Theoretical and Numerical

In some studies, there are theoretical and numerical approaches to investigate the effects of a dual-mass flywheel on vibration, including mathematical models that are created using only various methods [1, 9, 16–19]. For example, in [16] the linear approach is preferred by simplifying the model of DMF-CPVAs setup. Based on Lagrange's method, the equation of motion for the set-up has been derived. After analyzing the ability of the setup, it is observed that the utilization of CPVAs on the DMF is better on vibration isolation than damping vibrations at a specific frequency. The idle speed of a heavy vehicle was taken into consideration in [1, 9]. In these researches, the vehicle dynamics were determined with the help of the ADAMS program. In the new dual-mass flywheel structure, the springs are distributed on both sides of the damping disc, and the damping effect is examined with the help of simulation studies [1]. Idling natural frequency and idling natural vibration model were calculated by applying generalized Jacobian and MATLAB procedures. According to the results obtained, it is seen that DMF torsion absorber and power transfer torsion absorber are kept away from the idle condition when compared. Secondly, they form low-frequency torsions in the power transmission system. In a different study, analyses were performed on the segmented linear model, and the frequency of the model under sinusoidal response was obtained. When the studies and calculations were compiled, inertia, torsional stiffness and damping values were obtained. According to the analysis, when the inertia of both flywheels is increased, the vibration damping ability of dual mass flywheel increases [17]. The calculation of the natural frequency of the whole system is valid only if, the values such as vehicle inertia, transmission gear ratios and tire torsional stiffness are known. In order to reduce the acceleration of the secondary mass, it is provided by gear increase or reduction of the angular speed [18]. In another study, the friction between the primary flywheel of the dual mass flywheel and the arc spring has been investigated by Python program [19]. The simulations were made by combining Newmark and Newton methods. The accuracy of the model has also been checked in programs as AVL Excite. Columb friction model or inverse tangent functions were used to model the friction between two surfaces. According to the results, it is observed that if the friction and viscous damping values are low, it does not cause a significant problem at low engine speeds.

In addition, the dual-mass flywheels observed in the studies may also vary structurally. In addition to addressing the dual mass flywheel, mechanisms such as DMF-CPVAs – (Centrifugal Pendulum Vibration

Absorbers) [16], DMF-CS – (Circumferential Arc Spring Dual Mass Flywheel) [4, 17, 20] were also encountered.

Methods

NDSolve Solver

In Mathematica software, it is possible to find the solution of ODEs and PDEs numerically by NDSolve command. Instead of writing a function, it gives an appropriate interpolation function, namely, *InterpolatingFunction*. Boundary values may also be specified using Dirichlet Condition (Boundary condition type for a partial differential equation which gives the prescribed value of the function on a surface) and Neumann Condition (boundary condition type for a partial differential equation which gives the first derivative on a surface). The command may solve some of the differential-algebraic equation types, that either contains algebraic equations, differential equations or both of them in one equation. The iteration process is valid for NDSolve solution. In the first step of the iteration, a specific prescribed value is considered. Secondly, the output of each iteration is then the starting point of the next iteration again. Finally, this repetition process generates a sequence of output up to endpoint. To define the maximum number of steps of the iteration process, we can use "MaxSteps" option by selecting Automatic mode. In addition to this, *StartingStepSize*, *MaxStepSize*, and *NormFunction* items are used to describe (i) size of the step at the beginning, (ii) maximum size of step in independent variable of the equation and (iii) the norm of error estimation, respectively. If it is not given the maximum number of iteration for the process, Mathematica takes 10,000 as a stopping criterion. Given that tolerances are affected by error estimates. They can be scaled by combining the errors for different terms by satisfying the following condition.

$$f\left[\left\{\frac{error_1}{tolerance_r \,|x_1| + tolerance_a}, \frac{error_2}{tolerance_r \,|x_2| + tolerance_a}\right.\right.$$
$$\left.\left. \dots, \frac{error_n}{tolerance_r \,|x_n| + tolerance_a}\right\}\right] \leq 1 \tag{5.1}$$

where the function f represents the norm function that computing norms of error estimate in NDSolve solver, $error_i$ is the ith component of the error and x_i is the ith component of the current solution, n is the number of components. Absolute and relative tolerances are denoted by $tolerance_a$ and $tolerance_r$, respectively. In the procedure, an embedded error estimator tries to obtain an appropriate step size in an improved version of the explicit Runge-Kutta method, which is also adaptive embedded pairs of orders.

The options *TimeIntegration*, *BoundaryValues* and *EquationSimplification* can be adjusted by the user for NDSolve command. Depending on the DE type, these options correspond to systems of DE, Boundary Value Problems (BVPs) in ODE and simplified equation, respectively. *Adams*, *BDF*, *ExplicitRungeKutta*, *ImplicitRungeKutta* and *Symplectic-PartitionedRungeKutta* approaches are also hybridised with explicit Runge-Kutta method by time integration settings. The method starts with a trial step at the midpoint for the domain, and this leads to reduce lower-order error terms [21].

FindFit Solver

This solver is utilized to obtain the best-fitted function to the prescribed data numerically. *"ConjugateGradient"*, *"Gradient"*, *"LevenbergMarquardt"*, *"Newton"*, *"NMinimize"*, and *"QuasiNewton"* are the method options which can also be selected according to the given problem. In the present problem, the Levenberg-Marquardt method is selected as an appropriate process to calculate the regression coefficients, which are also a sub-problem of least-square approximation. It is a method for minimizing a sum-of-squares objective function. In this method, the following mathematical expression (Equation 5.2) is valid, and this corresponds to the variation of Gauss-Newton and Gradient Descent updates for the prescribed parameters.

$$[J^T W J + \lambda I]h = J^T W(y - \hat{y}) \tag{5.2}$$

In this equation, J^T, J, W are the transposed form of the Jacobian, traditional Jacobian and diagonal weighting matrix, respectively. λ represents the damping parameter, which can be adjusted to be large or small. I is the identity matrix, h is the perturbation, y is a measured point set, and the fitted function is denoted by $\hat{y}$ [22].

In addition to *NDSolve* and *FindFit* solver, another important command performed in the optimization part of the present design problem is *NMinimize*. A detailed explanation of *NMiminize* is given in the previous section (see Chapter 3).

Engineering Model

Components of a typical DMF are (1) primary flywheel, (2) spring, (3) flange and (4) secondary flywheel as shown in Figure 5.1.

Although the flange is a component of the dual mass flywheel, it is assumed as a single piece with the component which is the secondary flywheel. Accordingly, this model has six parameters. Firstly, the parameters whose J_1 and J_2 are the moment of inertias that belongs to the primary and secondary flywheel, respectively. The primary flywheel and secondary flywheel are merged using the torsional spring k and torsional damper

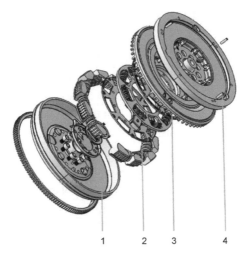

1   2   3   4

Figure 5.1 Dual mass flywheel.

c that determine the other parameters. The system has an engine torque, $Me(t)$, and this torque acts upon a DMF. A torque that, contrary to engine torque, namely counter-torque, is applied to the backside of driveline [23]. It is called Dual Mass Torsional Vibration Dynamic Absorber (DMTVDA) so that the system is more public than the traditional ones.

Mathematical Model

In order to obtain a mathematical model, the equations were constituted by considering the DMTVDA model and input parameters [23].

$$J_1\ddot{\varphi}_1 + c_1(\dot{\varphi}_1 - \dot{\varphi}_2) + k_1(\varphi_1 - \varphi_2) = M_e(t) \tag{5.3}$$

$$J_2\ddot{\varphi}_2 + c_1(\dot{\varphi}_2 - \dot{\varphi}_1) + k_1(\varphi_2 - \varphi_1) + c_2(\dot{\varphi}_2 - \dot{\varphi}_v) + k_2(\varphi_2 - \varphi_v) = 0 \tag{5.4}$$

where $c_2(\dot{\varphi}_2 - \dot{\varphi}_v) + k_2(\varphi_2 - \varphi_v)$ represents the gearbox input (the output side of the system) torque $M_v(t)$. All the parameters appear in Equations 5.3 and 5.4 and the definitions are listed in Table 5.1.

Some restrictions have been introduced to solve the design problem. Firstly, the six-cylinder truck engine will be taken into consideration. In this model which includes the DMF and output shaft joined to the gearbox, the following assumption is valid for engine torque as a mathematical expression [21].

$$M_e(t) = M_0 + M_1 * \sin(w_e t + \alpha_1) \tag{5.5}$$

where α_1 is phase angle, M_0 is constant torque, M_1 represents the amplitude of the wave. Lastly, it is assumed that the output side of the system

Table 5.1 Meaning of notations.

$M_e(t)$	Engine Torque
$\varphi_1(t)$	The absolute angle of rotation of the primary flywheel
$\dot{\varphi}_1(t)$	The absolute angular speed of rotation of the primary flywheel
$\ddot{\varphi}_1(t)$	The absolute angular acceleration of the primary flywheel
$\varphi_2(t)$	The absolute angle of rotation of the secondary flywheel
$\dot{\varphi}_2(t)$	The absolute angular speed of rotation of the secondary flywheel
$\ddot{\varphi}_2(t)$	The absolute angular acceleration of the secondary flywheel
$\varphi_v(t)$	The absolute angle of rotation of the output
$\dot{\varphi}_v$	The absolute angular speed of rotation of the output
k_1	Torsional stiffness coefficient of the primary flywheel
c_1	Torsional damping coefficient of the primary flywheel
k_2	Torsional stiffness coefficient of the secondary flywheel
c_2	Torsional damping coefficient of the secondary flywheel
$M_v(t)$	Output torque

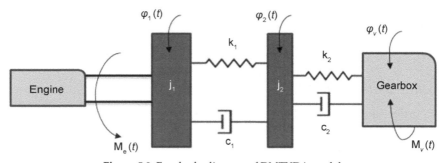

Figure 5.2 Free body diagram of DMTVDA model.

does not have any vibration. Therefore, the following equations will be appropriate.

$$\varphi_v = w_v t \text{ and } \dot{\varphi}_v = w_v \tag{5.6}$$

Three different cases (900, 1500, and 2,000 RPM) can be considered in the speed range used for truck engines. These values correspond to low, medium and high-speed models, respectively. Also, an engine that is a four-stroke six-cylinder will be preferred. One of the factors affecting the excitation frequency is the number of cylinders. The main excitation frequency matches up to the number of engines multiplied by half the number of cylinders [23].

Now, Equation 5.6 is necessary to calculate the engine torque. It should be known that the velocity which comes from the gearbox side is applied for this calculation. 1/3 times the engine speed can be taken for angular velocity from the gearbox side, and it is assumed that $\frac{W_e}{3} = W_v$. 900, 1,500 and 2,000 RPM also correspond to 15, 25 and 33.33 Hz, respectively. In this regards, engine torque should be calculated by the process mentioned above.

Based on the experiences in this field, the best values for M_1 and M_0 are advised as 300 and 500, respectively. According to these selections, the engine torques for three cases (low, medium and high speed) are obtained as follows [24].

$$M_{e1}(t) = 300 + 500 * \sin(45 * 2\pi t) \tag{5.7}$$

$$M_{e2}(t) = 300 + 500 * \sin(75 * 2\pi t) \tag{5.8}$$

$$M_{e3}(t) = 300 + 500 * \sin(100 * 2\pi t) \tag{5.9}$$

Optimization Problem Definition

In this study, in order to reduce the torsional vibration of the heavy vehicle's engine, four optimization problems have been defined (see Table 5.2). This work is focusing on optimization the DMF for different scenarios of the truck. The primary function is to isolate the transmission input from the vibration generated by the engine. The system parameters J_1, J_2, the primary stiffness coefficient k_1, secondary stiffness coefficient k_2 and the damping coefficients c_1, c_2 are considered as design variables.

A neuro-regression approach has been implemented for the present study. The resulting model is in the form of a second order rational polynomial as:

$$(a0 + a1j_1 + a10j_1k_1 + a11j_1 k_2 + a12 c_1j_2 + a13c_2j_2 + a14j_2 k_1 + a15j_2 k_2 + a16c_2 c_1$$
$$+ a17c_1k_1 + a18c_1k_2 + a19c_2k_1 + a2j_2 + a20c_2k_2 + a21k_1 k_2 + a22j_1^2 + a23j_2^2 + a24c_1^2$$
$$+ a25c_2^2 + a26k_1^2 + a27k_2^2 + a3c_1 + a4c_2 + a5k_1 + a6k_2 + a7j_1j_2 + a8c_1j_1 + a9c_2j_1)/$$
$$(b0 + b1j_1 + b10j_1 k_1 + b11j_1 k_2 + b12c_1j_2 + b13c_2 j_2 + b14j_2 k_1 + b15j_2 k_2 + b16c_2$$
$$c_1 + b17c_1 k_1 + b18c_1 k_2 + b19c_2 k_1 + b2j_2 + b20c_2 k_2 + b21k_1 k_2 + b22j_1^2 + b23j_2^2 +$$
$$b24c_1^2 + b25c_2^2 + b26k_1^2 + b27k_2^2 + b3c_1 + b4c_2 + b5k_1 + b6k_2 + b7j_1 j_2 + b8c_1 j_1 +$$
$$b9c_2j_1) \tag{5.10}$$

Results and Discussion

Considering only the low speed model (900 RPM), the system of equation is solved for 768 different matches (combination of the cases for $J_1 \in \{1.8, 1.7, 1.6, 1.5, 1.4, 1.3, 1.2, 1.1\}$, $J_2 \in \{0.6, 0.5\}$, $c_1 \in \{30, 25\}$, $c_2 \in \{15, 20\}$, $k_1 \in \{18,000, 16,000, 12,000, 20,000\}$, $k_2 \in \{16,500, 11,000, 5,500\}$). The results have been obtained as in an interpolation function representation.

Table 5.2 Four optimization scenarios.

Problem No.	Objectives	Constraints
Problem 1	Minimize $\varphi_1 (J_1, J_2, c_1, c_2, k_1, k_2) - \varphi_2 (J_1, J_2, c_1, c_2, k_1, k_2)$	$1.1 \leq J_1 \leq 1.8,$ $0.5 \leq J_2 \leq 0.6,$ $25 \leq c_1 \leq 30,$ $15 \leq c_2 \leq 20,$ $18{,}000 \leq k_1 \leq 20{,}000,$ $5{,}500 \leq k_2 \leq 16{,}500$
Problem 2	Minimize $\varphi_1 (J_1, J_2, c_1, c_2, k_1, k_2) - \varphi_2 (J_1, J_2, c_1, c_2, k_1, k_2)$	$1.1 \leq J_1 \leq 1.8$ $0.5 \leq J_2 \leq 0.6,$ $25 \leq c_1 \leq 30 \ \& \ c_1 \in \text{Integers}$ $15 \leq c_2 \leq 20 \ \& \ c_2 \in \text{Integers}$ $18{,}000 \leq k_1 \leq 20{,}000 \ \&$ $k_1 \in \text{Integers}$ $5{,}500 \leq k_2 \leq 16{,}500 \ \&$ $k_2 \in \text{Integers}$
Problem 3	Minimize $\varphi_1 (J_1, J_2, c_1, c_2, k_1, k_2) - \varphi_2 (J_1, J_2, c_1, c_2, k_1, k_2)$	$1.1 \leq J_1 \leq 1.8,$ $0.5 \leq J_2 \leq 0.6,$ $c_1 \in \{25, 30\}$ $c_2 \in \{15, 20\}$ $k_1 \in \{18{,}000, 18{,}500, 19{,}000, 19{,}500, 20{,}000\}$ $k_2 \in \{5{,}500, 6{,}500, \ldots, 15{,}500, 16{,}500\}$
Problem 4	Minimize $\varphi_1 (J_1, J_2, c_1, c_2, k_1, k_2) - \varphi_2 (J_1, J_2, c_1, c_2, k_1, k_2)$	$1.1 \leq J_1 \leq 1.8,$ $0.5 \leq J_2 \leq 0.6,$ $c_1 \in \{30, 25\}$ $c_2 \in \{15, 20\}$ $k_1 \in \{18{,}000, 16{,}000, 12{,}000, 20{,}000\}$ $k_2 \in \{16{,}500, 11{,}000, 5{,}500\}$

Table 5.3 represents the values of the numerical solutions of the randomly selected nine solutions. When the table is examined, it is seen that different φ_1 and φ_2 values are obtained at different times. Especially at t = 0.1s, it is observed that the standard deviation is higher than other time values. A deviation of 25% is found at the time of t = 0.1s, whereas at t = 0.5s, the highest deviation remains 0.45%. Lastly, similar behaviour is observed for t = 10s and t = 0.5s. Another remarkable point in Table 5.3 is for the values where φ_1 and φ_2 are highest, k_2 is 5,500 Nm/rad.

Before constituting the mathematical model given in the table below, 768 lines of data are divided into two parts as 20% and 80% of it. The multiple nonlinear regression model is constituted using 80% part of the data, and the values in the 20% ones are used for testing. In Table 5.4, the statistical accuracy of the model is shown, and the results are given for R^2_{Training} and R^2_{Testing} as 0.99218 and 0.991596, respectively. In the given model in Table 5.3 boundedness of the function (created model) considering the limitation of the engineering parameters have been checked. The maximum angular displacement is obtained as 0.068 radians (3.941 degrees) while

Table 5.3 Randomly selected differential equation system results for different input parameters.

Case Number	J_1 (kg.m²)	J_2 (kg.m²)	c_1 (Nms/rad)	c_2 (Nms/rad)	k_1 (Nm/rad)	k_2 (Nm/rad)	t=0.1 (s)		t=5 (s)		t=10 (s)	
							φ_1 (rad)	φ_2 (rad)	φ_1 (rad)	φ_2 (rad)	φ_1 (rad)	φ_2 (rad)
1	1.8	0.6	30	20	18,000	16,500	1.46257	1.47199	75.034	75.022	150.034	150.022
50	1.8	0.5	30	15	18,000	16,500	1.45312	1.46523	75.034	75.023	150.034	150.023
100	1.7	0.6	25	15	18,000	16,500	1.43566	1.45584	75.0338	75.0227	150.034	150.023
150	1.7	0.5	30	15	18,000	11,000	1.5669	1.53633	75.0429	75.032	150.043	150.032
200	1.6	0.6	25	15	18,000	11,000	1.55051	1.52785	75.0429	75.0309	150.043	150.031
300	1.5	0.6	25	15	18,000	5,500	1.79405	1.73224	75.0702	75.0576	150.07	150.058
500	1.3	0.6	25	15	16,000	11,000	1.47756	1.47768	75.0447	75.0311	150.045	150.058
700	1.1	0.6	25	15	12,000	16,500	1.41457	1.45209	75.0419	75.022	150.042	150.022
768	1.1	0.5	25	15	20,000	5,500	1.66225	1.62263	75.0664	75.0628	150.066	150.063

the minimum angular displacement is obtained as –0.0713549 radians (–4.088 degrees). As a result, it is confirmed that the model works correctly between these limits. Therefore, the model is to be realistic.

The optimum results are obtained by minimizing the differences between angular displacements of the first and second flywheels using values of the parameters in the aforementioned section. The optimization results are given in Table 5.5. As it is seen in the first optimization problem, the relative displacement $\varphi_1 - \varphi_2$ value is observed to be considerably less than 0.00263016 that is the smallest value obtained with the system of differential equation solution. When Problem 1 is examined, it can be seen that there are some values of parameters that cannot be used in the real application. However, it should be noted that the aim of problem one is to show the theoretical limits. For the second optimization problem, the value of $\varphi_1 - \varphi_2$ appears to be negative; therefore, the position of these parts (the first and second flywheels) is opposite to the relative displacement of

Table 5.4 Introduced model and its statistical accuracy.

Model ($\varphi_1 - \varphi_2$)	$R^2_{training}$	$R^2_{testing}$
$(-3.03513\times10^6+3.24015\times10^5c_1-5.9229\times10^3\,c_1{}^2+4.40512\times10^5c_2-2.97c_1c_2$ $-1.26171\times10^4c_2{}^2+8.65994\times10^4J_1-2.42361\times10^2c_1J_1-5.1525\times10^2c_2J_1$ $-1.54489\times10^4J_1{}^2-1.86464\times10^7J_2+8.2748\times10^3c_1J_2+5.05001\times10^3c_2J_2$ $+2.12179\times10^4J_1J_2+1.65423\times10^7J_2{}^2-9.03968k_1-41351\times10^{-7}k_1$ $-64250.7\times10^{-7}c_2k_1+6.53301\times10^{-1}J_1k_1-8.99026J_2k_1+4099.37\times$ $10^{-7}k_1{}^2-7.09892\times10^{-1}k_1k_2-919644\times10^{-7}c_1k_2-565448\times10^{-7}c_2k_2-888043$ $\times10^{-7}J_1k_2+9.37787\,J_2k_2+1054.8\times10^{-7}k_1k_2-372.274\times10^{-7}k_2{}^2)/$ $(2.086\times10^6+3.89261\times10^4c_2-2.29017\times10^3c_2{}^2+1.82744\times10^4c_2-$ $2.30952c_1c_2-1.84742\times10^3c_2{}^2+1.71559\times10^6J_1-6.59494x-1.87113x-$ $5.85142x-5.35259x+2.55628x+1.491\times10^5c_2J_2+5.72963\times10^5J_1J_2-$ $1.06301\times10^7J_2{}^2-3.14884\times10^7J_2{}^2+1.36021c_1k_1+1.3935c_2k_1+1.7277x1$ $10^2J_1k_1-4.28398\times10^2J_2k_1+87638.6\times10^{-7}k_1{}^2-8.72722\times10k_2-2.82216c_1k_2$ $-1.59647c_2k_2-1.38122J_1k_2+3.6142\times10^2J_2k_2+49658.6\times10^{-7}k_1k_2-$ $13831.3\times10^{-7}k_2{}^2)$	0.99218	0.991596

Table 5.5 Optimization results for different scenarios.

	$(\varphi_1 - \varphi_2)$ (rad)	J_1 (kg . m²)	J_2 (kg . m²)	c_1 (Nms/ rad)	c_2 (Nms/ rad)	k_1 (Nm/ rad)	k_2 (Nm/ rad)
Optimization Problem 1	0.000000010	1.10009	0.507709	27.3077	19.9999	19,999.8	5,500
Optimization Problem 2	–0.000155602	1.79996	0.599997	28	19	18,001	5,501
Optimization Problem 3	0.00269078	1.1	0.6	25	15	20,000	16,500
Optimization Problem 4	0.00269078	1.1	0.6	25	15	20,000	16,500

those of the other problems. The third and fourth optimization problems are studied that limited parameters value ranges in order that it can be used in the real application. According to the results of optimizations, it is seen that the obtained minimum $\varphi_1 - \varphi_2$ value is very close to the smallest value obtained by using the system of a differential equation. It is technically possible that J_1 and J_2 are continuous; therefore, discrete values are not used in the constraint definition. The best improvement within the theoretical limits has been observed to be 99%.

Conclusion

In this study, dual mass flywheel parameter optimizations have been performed for transferring vibration that are generated in the engine of a heavy vehicle to the driveline of the vehicle with the lowest amplitude. A differential equation system is solved by explicit Runge-Kutta for different input parameters values, and then a data set is obtained using full factorial design. The model is constituted using these data values by the Levenberg-Marquardt method in Mathematica program. The improvements resulting from optimizations have been observed to be 99%. The results show that optimization plays a significant role in DMF designs, and the results obtained in the Mathematica program provide an appropriate methodological way for these to be developed. From this point of view, the relevant parameters must have a higher number of different values in order to reach the efficiency increase obtained with theoretical limits.

References

[1] Chen, D.M., Shi, X.F. and Liu, G.Q. 2012. Design and simulation study of a new type of dual mass flywheel. Appl. Mech. Mater. 184-185: 130–133.

[2] Theodossiades, S., Gnanakumarr, M., Rahnejat, H. and Kelly, P. 2006. Effect of a dual-mass flywheel on the impact-induced noise in vehicular powertrain systems. Proc. Inst. Mech. Eng. D J. Automob. Eng. 220: 747–761.

[3] Zeng, L.P., Song, L.Q. and Zhou, J.P. 2015. Design and elastic contact analysis of a friction bearing with shape constraint for promoting the torque characteristics of a dual mass flywheel. Mech. Mach. Theory 97: 356–374.

[4] Wu, C., Peng, Y.F., Xiang, C., Mao, X.Y., Hu, J.H. and Liu, Y.O. 2015. Static torsional stiffness computation of circumferential arc spring dual mass flywheel. Adv. Mater. Res. 1065-1069: 2080–2085.

[5] Wang, Y., Qin, X., Huang, S. and Deng, S. 2016. Design and analysis of a multi-stage torsional stiffness dual mass flywheel based on vibration control. Appl. Acoust. 104: 172–181.

[6] Song, L.Q., Zeng, L.P., Zhang, S.P., Zhou, J.D. and Niu, H.E. 2014. Design and analysis of a dual mass flywheel with continuously variable stiffness based on compensation principle. Mech. Mach. Theory 79: 124–140.

[7] Kim, T.H., Song, H.L., Hwang, S.H. and Kim, H.S. 2006. Analysis of dual mass flywheel using discrete arcspring model. Key Eng. Mater. 326-328: 1607–1610.

[8] Schaper, U. Sawodny, O., Mall, T. and Blessing, V. 2009. Modeling and torque estimation of an automotive dual mass flywheel. The Proceedings of the American Control Conference, St. Louis, MO, USA.

[9] Meng, X. and Li, J. 2011. Idling natural characteristic analysis of the torsion absorber with dual mass flywheel. Appl. Mech. Mater. 80-81: 860–864.

[10] Chen, L., Zeng, R. and Jiang, Z. 2015. Nonlinear dynamical model of an automotive dual mass flywheel. Adv. Mech. Eng. 7(6): 1–11.

[11] Mahl, T. and Sawodny, O. 2010. Modelling of an automotive dual mass flywheel. 5th IFAC Symposium on Mechatronic Systems, Marriott Boston Cambridge, MA, USA.

[12] Ozansoy, O., Tevruz, T. and Mugan, A. 2015. Multiobjective pareto optimal design of a clutch system. Int. J. Eng. Tech. 1(1): 26–43.

[13] Yang, X., Zhang, T-S. and Lei, N-L. 2017. Experimental study on torsional vibration of transmission system under engine excitation. The Proceedings of the 3rd International Conference, Applied Mechanics and Mechanical Automation.

[14] Chen, D., Xu, J.H., Shi, X.F. and Ma, Y.Y. 2012. Experimental study on torsional vibration of dual mass flywheel. Adv. Mater. Res. 490-495: 2318–2322.

[15] Kang, T.S., Kauh, S.K. and Hua, K.P. 2009. Development of the displacement measuring system for a dual mass flywheel in a vehicle. Proc. Inst. Mech. Eng. DJ. Automob. Eng. 223: 1273.

[16] Zhao, G., Chen, L. and Jiang, Z. 2013. Linear analysis for performance of dual mass flywheel with centrifugal pendulum vibration absorbers system. Telkomnika. 11(5): 2371–2376.

[17] Zeng, R., Chen, L. and Jiang, Z. 2015. Frequency response analysis of damped dual mass flywheel. Appl. Mech. Mater. 724: 271–274.

[18] Pavlov, N. 2018. Numerical simulation on the vibration of a test bed with engine with dual mass flywheel. International Scientific Journal, Machines Technologies, Materials, pp. 49–52.

[19] Johansson, D. and Karlsson, K. 2017. Simulation Models of Dual Mass Flywheels. M.S. Thesis, Chalmers University of Technology, Gothenburg, Sweden.

[20] Jiang, Z.F. and Chen, L. 2010. Research on the method of circumferential spring dual mass flywheel damper matching with diesel engine. Key Eng. Mater. 419-420: 65–68.

[21] Sofroniou, M. and Knapp, R. 2008. AY-GÜN Advanced Numerical Differential Equation Solving In Mathematica. Wolfram Mathematica Tutorial Collection. Retrieved from http:\\ library.wolfram.com.

[22] Wellin, P.R., Gaylord, R.J. and Kamin, S.N. 2005. An Introduction to Programming with Mathematica. Cambridge University Press, New York, United States of America.

[23] Rahnejat, H. 2010. Tribology and Dynamics of Engine and Powertrain: Fundamentals, Applications and Future Trends. Woodhead Publishing Limited, Sawston, Cambridge.

[24] Bourgois, G. 2016. Dual Mass Flywheel for Torsional Vibrations Damping Parametric Study for Application in Heavy Vehicle. M.S. Thesis, Chalmers University of Technology, Gothenburg, Sweden.

CHAPTER 6

Wind Turbine Optimization by Using Stochastic Methods

Ebru Düz,[1] *Levent Aydin,*[2] *Selda Oterkus*[3,*] *and Mahir Tosun*[4]

Introduction

Wind energy is a more suitable alternative energy source than most of the others. Wind turbines are one of the cleanest energy sources. Therefore, it is of great importance in reducing fossil fuel dependence. Wind energy appears as a notable contributor to the overall energy supply mix. Onshore and offshore, wind farms will grow further and surround increasingly bigger areas. Wind energy technology is used to reduce the cost of wind energy to produce energy at a more competitive price. Even during the global recession and the financial crisis, the wind energy industry is continually getting bigger. Recently, wind power has been shown to be the world's fastest-growing renewable energy source.

Wind turbine systems have been developed to produce power with constant angular velocity and force [1]. It provides the production of wind energy with the aerodynamic force developing on the blades of wind

[1] Dokuz Eylül University, The Graduate School of Natural and Applied Sciences, İzmir, Turkey.
Email: ebruduz0@gmail.com
[2] İzmir Katip Çelebi University, Department of Mechanical Engineering, İzmir, Turkey.
Email: leventaydinn@gmail.com
[3] University of Strathclyde, PeriDynamics Research Centre, Department of Naval Architecture, Ocean & Marine Engineering, Glasgow, Scotland.
[4] Borusan EnBW Energy, Wind and Solar Engineering Manager, İstanbul, Turkey.
Email: mahirtosun@gmail.com
* Corresponding author: selda.oterkus@strath.ac.uk

turbines [2]. However, the wind velocity is different in magnitude and direction. Besides, the velocity is not the same at varying heights for a massive rotor [1].

Harsh environmental conditions can cause failures in different parts of wind turbines, such as main bearings, gearboxes, and generators. Besides, traffic shortages and loss of downtime have a significant impact on energy costs. Therefore, the rapid development of the wind power industry requires higher performance and reliability for the equipment. The actual use of wind turbines may reduce if reliable operations are not provided. Increase in operating and maintenance (O&M) costs may also reduce the economic benefit of wind power [3].

One of the main issues for the design of control systems of wind turbines is to maximize wind power generation. The unstable aerodynamic loads on a blade should be considered in the design process of a wind turbine blade to ensure the reliability of the wind turbine system. Optimization methods are commonly used in the design process of wind turbine blades and other system components [4]. Fuglsang and Madsen [5] developed a design optimization method for the design of the horizontal axis wind turbine (HAWT) to minimize the cost of energy with the constraints of aerodynamic load and noise. Benini and Toffolo [6], using a multi-purpose optimization of the design of HAWTs, provided a performance change between annual energy generation and energy cost objectives. Jurezko et al. [7] developed a computer program package to optimize the wind turbine blades for the highest possible power output for various aerodynamic loads. Casas et al. [8] created an automated design environment combined with an aerodynamic simulation. Using the design optimization process, they have improved the aerodynamic performance of a wind turbine.

Increasing the efficiency of a wind turbine produces more power, which reduces the need for expensive power generation. Since the seventh century, people have been using the wind to generate energy. The usage of renewable energy requires technologies that harness natural phenomena, such as sunlight, wind, waves, water flow, biological hydrogen production, and geothermal heat. Among the other energy sources, much progress has been made in technology to exploit wind. Energy transmitted by wind depends on the swept area of the rotor, speed of the wind, and the density of the air. The blade is the critical component in capturing wind energy. It plays a vital role in the entire wind turbine.

The interaction between the rotor and the wind affects the power generation of the turbine [9]. Many studies in the literature take into account the wind conditions of the site with regards to the wind turbine design. Some of the studies are related to a single criterion optimization method, where the objective function is the cost of generated electrical energy to maximize the measured energy at the lowest cost. Some of the design optimization methods are based on a single criterion where the objective function is the cost of generated electrical energy and to maximize the measured energy at the lowest cost.

Applications of multi-level design configuration for the wind turbine system (WTS) is also available in the literature [5, 10]. Fuglsang et al. [5] minimized energy costs by changing rotor parameters and blade shape. They performed a multidisciplinary optimization study by taking into account power generation, structural loading, noise emission, lifetime, and reliability. A similar multi-level approach is used for optimal rotor configuration in WTS [10]. This multi-level approach involves two procedures. The first is optimizing blade geometry to maximize annual energy production. The second is the structural blade design that minimizes the bending movement at the blade root.

There are many studies in the literature related to the optimization of parameters affecting the power of a wind turbine. Sargolzaei et al. [11] used artificial neural networks (ANNs) to estimate the power factor and torque of wind turbines based on experimental data collected from seven prototype vertical rotors tested in a wind tunnel. Abdel-Aal et al. [12] argued that predicting wind speed is very important for the operating and protection of wind farms and their economic concretion into power grids. They argued that it is also essential for many forms of implementation used in aviation, shipping, and the environment. Although modern machine learning techniques are used in conjunction with neural networks, they have proved difficult to produce significant improvements in the performance of a simple persistence model.

Monfared et al. [13] proposed a new fuzzy logic and ANN-based approach for wind speed estimation. New approaches to fuzzy logic have provided a much less rule based and also provided better precision than the traditional method for wind speed estimation. The experimental results support that the suggested method would supply a low calculation time as well as superior wind speed prediction performance.

Bharanikumar et al. [14] have attempted to present a Maximum Power Point Tracking (MPPT) control algorithm for a variable speed wind turbine driven permanent magnet generator. The efficiency of the Wind Energy Conversion System was maximized by operating the wind turbine generator in such a way that the rotor speed is proportional to the wind speed. MPPT algorithm was used to monitor the maximum power point for each speed value.

The main objective of this study is to determine the optimum design of a wind turbine structure by using mathematical models. The accuracy of the constructed models is checked through R^2 training and testing values. After testing the reliability of the models, it is aimed to maximize the generator torque by using Mathematica. Stochastic methods such as

Differential Evolution (DE), Random Search (RS) and Simulated Annealing (SA) and Nelder-Mead (NM) algorithms are used.

Problem Definition

In this study, a 5 MW wind turbine with a rotor diameter of 154 m, rotating speed of 11.3 rpm is selected. This wind turbine has three blades, its cut-in, cut-out, and rated wind speed parameters are 3, 25, 11.4 m/s, respectively [3].

Jin et al. [3] used the backpropagation neural network to predict the dynamic behavior of the 5 MW wind turbine. Back Propagation (BP) network is the most commonly used and most accomplished neutral network. A typical BP network mostly contains input, output, and hidden layers. The process is divided into forwarding information propagation and backward error propagation. The number of neurons at the input and hidden layers are found through testing. Similar to Jin et al. [3] generator torque (GT), generator speed (GS), blade tip deflection (BTD) and tower top deflection (TTD) are considered as the output variables, whereas, wind speed (w) and yaw angle (y) are considered as input variables.

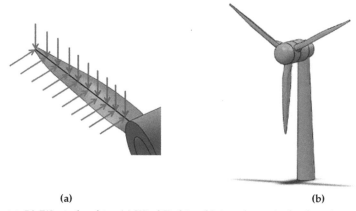

(a)   (b)

Figure 6.1 5 MW wind turbine (a) Wind Turbine (b) Aerodynamics loads acting on blades of a wind turbine [3].

As a statistical tool, regression analysis is an appropriate selection for estimating the relationships among the parameters which affect the engineering process. In the literature, many types of regression models are available such as Linear, Logistic, Nonlinear, and Stepwise to describe the phenomena. If the number of inputs is more than one, multiple regression procedures are used for the considered data. Furthermore, if the physical phenomenon is not linear in terms of parameters, nonlinear multiple regression models can be used to predict the phenomenological response of the process. Therefore, in this study, multiple nonlinear regression analysis is used [15].

For this design problem following steps are considered;

Step 1: Neuro-regression modeling

Step 2: Boundedness of the model

Step 3: Optimization

In Neuro-regression modeling step, 55 different datasets for generator speed, generator torque, blade tip deflection, and tower top deflection, wind speed, yaw angle [3] are divided into training and testing data. Training data consists of 80% of the data as provided in Table 6.1, and testing data is obtained using 20% of the data as provided in Table 6.2. The testing data is used to check the accuracy of the constructed model.

After separating the test and training data, ten different models are proposed, as shown in Table 6.3. The terms $a_0, a_1, ..., b_0, b_1$ are the arbitrary constants which are obtained by regression analysis. The calculated R^2 training, R^2 training-adjusted, R^2 testing, AIC (Akaike Information Criteria) and BIC (Bayesci Information Criteria) values for each model are presented in Table 6.4. The expressions used in calculations of R^2, R^2 training, R^2 training-adjusted, R^2 testing, AIC, and BIC are given in Chapter 1 in detail. Note that the column "standard" in Table 6.4 represents R^2 values obtained from all data.

In the second step, maximum and minimum values for the design variables (GS, GT, BTD, TTD) are calculated by using the proposed models in Table 6.3 to check whether the models are bounded or not. It is concluded that SONR (for BTD and TTD) and TONR (for GS and GT) models are appropriate for this engineering problem (see Table 6.5). The maximum and minimum values for the design variables are calculated by using the forms provided in Table 6.7.

In the third step, the engineering design problem is investigated by introducing two different optimization problems, which include maximizing generator torque with appropriate constraints (see Table 6.6). Generator torque in continuous and discrete domains are maximized based on DE, RS, SA, and NM algorithms for these two problems.

It is concluded that SONR (for BTD and TTD) and TONR (for GS and GT) regression models are suitable for this optimization problem. The form of these models and models with coefficient are shown in Table 6.7. The optimum values for the generator torque are obtained when the wind speed is between 18 and 24, and the yaw angle is between 14 and 40 (see Table 6.8).

Table 6.1 Training data [3].

INPUT		OUTPUT			
Wind Speed (m/s)	Yaw Angle (°)	Generator Speed (rpm)	Generator Torque (kNm)	Blade Tip Deflection (m)	Tower Top Deflection (m)
4	0	695.73	2.48	1.3	0.056
4	20	691.59	2.08	1.21	0.05
4	40	681.4	1.1	0.98	0.032
4	60	671.77	0.17	0.76	0.011
4	80	597.28	0	0.58	−0.0055
6	0	767.97	9.46	2.12	0.13
6	20	753.71	8.08	1.94	0.11
6	40	718.55	4.69	1.5	0.075
6	80	615.83	0	0.61	−0.0031
8	0	880.19	19.82	3.16	0.21
8	20	849.91	17.37	2.87	0.19
8	40	781.27	10.74	2.16	0.13
8	60	706.76	3.55	1.28	0.055
8	80	642.9	0	0.67	0.00013
10	0	1097.85	30.83	4.63	0.34
10	20	1039.51	27.64	4.17	0.3
10	40	865.23	18.85	2.93	0.19
10	60	741.7	6.92	1.67	0.086
12	0	1174.27	43.07	4.53	0.35
12	20	1158.41	42.06	5.39	0.41
12	60	787.06	11.3	2.11	0.12
12	80	674.88	0.47	0.76	0.0081
14	0	1173.94	43.08	3.12	0.27
14	40	1150.46	38.79	5.07	0.37
14	60	840.43	16.45	2.59	0.15
16	40	1174.22	43.07	3.95	0.31
16	60	927.14	21.99	3.18	0.2
18	0	1173.77	43.09	1.7	0.22
18	20	1173.86	43.09	1.95	0.23
18	60	1047.91	28.09	3.93	0.26

Table 6.1 contd. ...

... Table 6.1 contd.

INPUT		OUTPUT			
Wind Speed (m/s)	Yaw Angle (°)	Generator Speed (rpm)	Generator Torque (kNm)	Blade Tip Deflection (m)	Tower Top Deflection (m)
20	20	1173.78	43.09	1.47	0.21
20	40	1174	43.08	2.41	0.25
20	80	708.58	3.72	1.11	0.032
22	20	1173.69	43.09	1.06	0.2
22	40	1174.03	43.08	1.91	0.23
22	80	721.1	4.93	1.21	0.039
24	20	1173.72	43.09	0.7	0.2
24	40	1173.97	43.08	1.49	0.21
24	60	1174.89	43.05	3.68	0.27
24	80	732.32	6.02	1.34	0.045

Table 6.2 Testing data [3].

INPUT		OUTPUT			
Wind Speed (m/s)	Yaw Angle (°)	Generator Speed (rpm)	Generator Torque (kNm)	Blade Tip Deflection (m)	Tower Top Deflection (m)
14	20	1174	43.08	3.51	0.29
18	80	697.95	2.7	1.01	0.026
10	80	670.46	0.052	0.72	0.004
12	40	1036.21	27.46	4.05	0.28
18	40	1173.95	43.08	3.05	0.27
14	80	680.81	1.04	0.83	0.014
24	0	1174.02	43.08	0.52	0.18
16	20	1173.91	43.08	2.61	0.25
6	60	684.13	1.36	0.95	0.03
20	0	1173.69	43.09	1.24	0.2
22	60	1156.99	41.48	5.06	0.34
16	0	1173.85	43.09	2.31	0.24
20	60	1140.82	34.82	4.6	0.31
16	80	688.5	1.79	0.92	0.019
22	0	1173.75	43.09	0.85	0.19

Table 6.3 Linear and non-linear regression model forms for the outputs (GT, GS, BTD, TTD) of the engineering system [16].

Model Name	Abbreviation	Formula
Multiple Linear Rational	LR	$GT = \dfrac{(a_0 + a_1 w + a_2 y)}{(b_0 + b_1 w + b_2 y)}$
Second-Order Multiple Non-Linear	SON	$GT = a_0 + a_1 w + a_2\, y + a_3\, w^2 + a_4\, y^2 + a_5\, wy$
Second Order Multiple Non-Linear Rational	SONR	$GT = \dfrac{(a_0 + a_2 y + a_4 y^2 + a_1 w + a_3 yw + a_5 w^2)}{(b_0 + b_2 y + b_4 y^2 + b_1 w + b_3 yw + b_5 w^2)}$
Third Order Multiple Non-Linear	TON	$GT = a_0 + a_1 w + a_2 y + a_3 w^2 + a_4 y^2 + a_5 wy + a_6 w^3 + a_7 y^3 + a_8 w^2 y + a_9 wy^2$
Third Order Multiple Non-Linear Rational	TONR	$GT = \dfrac{\begin{array}{c}(a_0 + a_1 w + a_2 y + a_3 w^2 + a_4 y^2 + a_5 wy \\ + a_6 w^3 + a_7 y^3 + a_8 w^2 y + a_9 wy^2)\end{array}}{\begin{array}{c}(b_0 + b_1 w + b_2 y + b_3 w^2 + b_4 y^2 + b_5 wy \\ + b_6 w^3 + b_7 y^3 + b_8 w^2 y + b_9 wy^2)\end{array}}$
Fourth Order Multiple Non-Linear	FON	$GT = a_0 + a_1 w + a_2 w^2 + a_3 w^3 + a_4 w^4 + a_5 y + a_6 wy + a_7 w^2\, y + a_8 w^3\, y + a_9\, y^2 + a_{10} wy^2 + a_{11}\, w^2 y^2 + a_{12} y^3 + a_{13} wy^3 + a_{14}\, y^4$
Fourth Order Multiple Non-Linear Rational	FONR	$GT = \dfrac{\begin{array}{c}(a_0 + a_1 w + a_2 w^2 + a_3 w^3 + a_4 w^4 + a_5 y + a_6 wy + a_7 w^2 y \\ + a_8 w^3 y + a_9 y^2 + a_{10} wy^2 + a_{11} w^2 y^2 + a_{12} y^3 \\ + a_{13} wy^3 + a_{14} y^4)\end{array}}{\begin{array}{c}(b_0 + b_1 w + b_2 w^2 + b_3 w^3 + b^4 w^4 + b_5 y + b_6 wy + b_7 w^2 y \\ + b_8 w^3 y + b_9 y^2 + b_{10} wy^2 + b_{11} w^2\, y^2 + b_{12} y^3 \\ + b_{13} wy^3 + b_{14} y^4)\end{array}}$
First Order Trigonometric Non-Linear	FOTN	$GT = a_0 + a_1 sinw + a_2 siny + a_3 cosw + a_4 cosy$
Second Order Trigonometric Non-Linear	SOTN	$GT = a_0 + a_1 sinw + a_2 siny + a_3 cosw + a_4 cosy + a_5 sin^2 w + a_6\, sin^2 y + a_7 cos^2 w + a_8 cos^2 y$
Second Order Trigonometric Non-Linear Rational	SOTNR	$GT = \dfrac{\begin{array}{c}(a_0 + a_1 sinw + a_2 siny + a_3 cosw + a_4 cosy \\ + a_5 sin^2 w + a_6 sin^2 y + a_7 cos^2 w + a_8 cos^2 y)\end{array}}{\begin{array}{c}(b_0 + b_1 sinw + b_2 siny + b_3 cosw + b_4 cosy \\ + b_5 sin^2 w + b_6 sin^2 y + b_7 cos^2 w + b_8 cos^2 y)\end{array}}$

Table 6.4 R^2, Adjusted R^2, AIC, and BIC values, calculated for four different outputs (GS: Generator Speed, GT: Generator Torque, BTD: Blade Tip Deflection, TTD: Tower Top Deflection).

Model	Output	R^2			Adjusted R^2			AIC		BIC	
		Standard	Training	Testing	Standard	Training	Testing	Standard	Training	Standard	Training
LR	GS	0.9878	0.9611	-0.243	0.9863	0.9542	-0.266	683.362	545.333	697.414	557.155
	GT	0.8223	0.7977	0.5646	0.8005	0.7620	0.5563	446.466	328.942	460.517	340.765
	BTD	0.7892	0.8081	-0.028	0.7634	0.7742	-0.048	189.671	137.734	203.722	149.556
	TTD	0.8714	0.8546	0.3035	0.8557	0.829	0.2903	-113.01	-78.543	-103.96	-66.721
SON	GS	0.9938	0.9941	0.7847	0.9931	0.9930	0.7806	645.809	469.912	659.86	481.735
	GT	0.9550	0.9550	0.7765	0.9495	0.9470	0.7722	370.899	268.8	384.951	280.622
	BTD	0.8732	0.8904	0.3741	0.8577	0.8711	0.3622	161.693	115.318	175.744	127.14
	TTD	0.9102	0.9136	0.5919	0.8992	0.8984	0.5842	-137.77	-99.367	-123.71	-87.545
SONR	GS	0.9989	0.9989	0.9585	0.9986	0.9985	0.9577	559.251	411.621	585.346	433.577
	GT	0.9949	0.9941	0.9789	0.9935	0.9916	0.9785	262.185	199.081	288.28	221.037
	BTD	0.9542	0.9749	0.5268	0.9414	0.9642	0.5178	117.653	68.249	143.748	90.205
	TTD	0.9801	0.9857	0.8047	0.9746	0.9796	0.8010	-208.75	-159.46	-182.66	-137.50
TON	GS	0.9973	0.9976	0.8238	0.9967	0.9969	0.8204	606.974	440.655	629.028	459.232
	GT	0.9851	0.9867	0.8332	0.9818	0.9823	0.8300	317.979	227.818	340.059	246.396
	BTD	0.9206	0.9513	-0.271	0.9030	0.9350	-0.294	143.954	90.881	166.034	109.459
	TTD	0.9481	0.9639	0.2109	0.9366	0.9519	0.1960	-159.95	-126.27	-137.87	-107.69
TONR	GS	0.9998	0.9998	0.9909	0.9997	0.9997	0.9907	469.53	343.903	511.684	379.37
	GT	0.9991	0.9993	0.9888	0.9985	0.9986	0.9885	183.327	128.183	225.481	163.649
	BTD	0.9969	0.9983	0.8947	0.9952	0.9966	0.8927	-15.247	-23.630	26.906	11.835
	TTD	0.9961	0.9966	0.5708	0.9938	0.9932	0.5627	-282.25	-201.23	-240.10	-165.76

Table 6.4 contd. ...

... Table 6.4 contd.

Model	Output	R²			Adjusted R²			AIC		BIC	
		Standard	Training	Testing	Standard	Training	Testing	Standard	Training	Standard	Training
FON	GS	0.9656	0.9682	-0.195	0.9636	0.9656	-0.217	734.494	531.309	742.523	538.065
	GT	0.9918	0.9924	0.824	0.9888	0.9879	0.8206	294.932	215.318	327.049	342.34
	BTD	0.9631	0.9718	0.507	0.9493	0.9550	0.4977	111.765	78.903	143.883	105.926
	TTD	0.9729	0.9770	0.6212	0.9628	0.9632	0.6140	-185.79	-134.38	-153.68	-107.36
FONR	GS	0.9999	0.9999	0.9971	0.9999	0.9999	0.9970	378.271	260.603	440.498	312.958
	GT	0.9998	0.9999	0.9647	0.9997	0.9999	0.9640	85.625	14.438	147.853	66.793
	BTD	0.9992	0.9994	-2.755	0.9984	0.9977	-2.826	-74.666	-47.057	-12.439	5.297
	TTD	0.9997	0.9998	-5.476	0.9993	0.9995	-5.599	-405.33	-318.97	-343.10	-266.62
FOTN	GS	0.9654	0.9640	0.2148	0.9620	0.9588	0.1999	738.784	540.269	750.828	550.402
	GT	0.7509	0.7225	0.2509	0.7260	0.6829	0.2367	463.09	339.581	475.09	349.714
	BTD	0.7644	0.7654	0.1560	0.7408	0.7319	0.1400	193.797	143.766	205.841	153.9
	TTD	0.7880	0.7733	0.3813	0.7668	0.7409	0.3696	-92.501	-62.754	-80.457	-52.620
SOTN	GS	0.9699	0.9666	0.3723	0.9640	0.9569	0.3604	739.167	545.269	759.24	562.157
	GT	0.7801	0.7389	0.389	0.7371	0.6631	0.3774	464.176	345.149	484.249	362.038
	BTD	0.7922	0.8062	0.1070	0.7516	0.75	0.0901	194.876	144.125	214.95	161.014
	TTD	0.8249	0.8081	0.5187	0.7906	0.7525	0.5096	-95.018	-61.439	-74.945	-44.550
SOTNR	GS	0.9732	0.9740	0.1066	0.9602	0.9528	0.0897	750.688	553.12	788.827	585.209
	GT	0.8276	0.833	0.3935	0.7438	0.6963	0.3820	468.78	345.279	506.919	377.368
	BTD	0.8253	0.8637	-0.037	0.7404	0.7523	-0.056	203.333	148.037	241.472	180.126
	TTD	0.7986	0.893	-15.25	0.7006	0.805	-15.55	-69.323	-66.786	-31.184	-34.697

Table 6.5 Boundedness check for the models.

	Maximum Predicted Value from the Model	Maximum Value from the Data	Minimum Predicted Value from the Model	Minimum Value from the Data
TONR model for GS	1208.4	1174.89	697.56	597.28
TONR model for GT	46.69	43.09	2.66	0
SONR model for BTD	5.48	5.39	1.26	0,52
SONR model for TTD	0.38	0.41	0.03	−0.005

Table 6.6 Optimization problems.

Problem No.	Problem Definition
1	**Maximize** Generator Torque **Subjected to** 4 < w < 24, 0 < y < 80, 597.28 < GS < 1174.89, −0.005 < TTD < 0.3974
2	**Maximize** Generator Torque **Subjected to** w ∈ {4, 6, 8, 10, 12, 14, 16, 18, 20, 22, 24}, y ∈ {0, 20, 40, 60, 80}, 597.28 < GS < 1174.89, 0.52 < BTD < 5.39, −0.005 < TTD < 0.3974

Table 6.7 Selected model forms and obtained coefficients for the outputs GS, GT, BTD, and TTD.

Output	Model Form	Model
GS	$\dfrac{(a_0 + a_1 w + a_2 y + a_3 w^2 + a_4 y^2 + a_5 wy + a_6 w^3 + a_7 y^3 + a_8 w^2 y + a_9 wy^2)}{(b_0 + b_1 w + b_2 y + b_3 w^2 + b_4 y^2 + b_5 wy + b_6 w^3 + b_7 y^3 + b_8 w^2 y + b_9 wy^2)}$	$(-398.43\ w^3 - 830.054\ w^2 y + 113258.\ w^2 + 18.5344\ wy^2\ \ 4230.26\ wy - 1.60858\ 10^6\ w + 5.20488\ y^3 + 1234.61\ y^2 + 65516.\ y + 7.85288\ 10^6)/$ $(-1.48283\ w^3 - 0.356374\ w^2 y + 150.663\ w^2 - 0.0948685\ wy^2 - 11.0526\ wy - 2229.72\ w + 0.0184229\ y^3 + 2.01038\ y^2 + 95.7296\ y + 11198.)$
GT	$\dfrac{(a_0 + a_1 w + a_2 y + a_3 w^2 + a_4 y^2 + a_5 wy + a_6 w^3 + a_7 y^3 + a_8 w^2 y + a_9 wy^2)}{(b_0 + b_1 w + b_2 y + b_3 w^2 + b_4 y^2 + b_5 wy + b_6 w^3 + b_7 y^3 + b_8 w^2 y + b_9 wy^2)}$	$(1.38565\ w^3 - 1.5735\ w^2 y + 32.5063\ w^2 + 0.0356132\ wy^2 - 9.52318\ wy + 2487.41\ w + -0.0198773\ y^3 - 0.0749288\ y^2 + 88.8451\ y - 6367.64)/$ $(-0.622138\ w^3 + 0.0589192\ w^2 y + 35.2628\ w^2 - 0.0360904\ wy^2 - 2.37164\ wy - 538.004\ w + 0.00408397\ y^3 + 0.419051\ y^2 + 12.304\ y + 3199.47)$

Table 6.7 contd. ...

...Table 6.7 contd.

Output	Model Form	Model
BTD	$\dfrac{(a_0 + a_2y + a_4y^2 + a_1w + a_3yw + a_5w^2)}{(b_0 + b_2y + b_4y^2 + b_1w + b_3yw + b_5w^2)}$	$(11.5075\ w^2 - 1.49638\ wy - 243.28\ w$ $-0.679497\ y^2 + 68.9469\ y + 2242.96)/$ $(17.8564\ w^2 - 3.35706\ wy - 380.88\ w +$ $0.105935\ y^2 + 43.1297\ y + 2223.42)$
TTD	$\dfrac{(a_0 + a_2y + a_4y^2 + a_1w + a_3yw + a_5w^2)}{(b_0 + b_2y + b_4y^2 + b_1w + b_3yw + b_5w^2)}$	$(0.0297636\ w^2 - 0.0174438\ wy +$ $0.611894\ w - 0.00268026\ y^2 + 0.291031$ $y - 1.76271)/(0.758829\ w^2 - 0.217027$ $wy - 12.6522\ w + 0.0110345\ y^2 + 2.24272$ $y + 73.0047)$

Table 6.8 Optimum results.

	INPUT			OUTPUT			
Problem No.	Wind Speed (m/s)	Yaw Angle (°)	Generator Speed (rpm)	Generator Torque (kNm)	Blade Tip Deflection (m)	Tower Top Deflection (m)	Opt. Method
1	24	39.805	1174.89	44.1475	1.5993	0.1959	NM
	16.852	14.763	1174.89	44.6829	2.7248	0.2882	DE
	16.852	14.766	1174.89	44.6829	2.7249	0.2882	SA
	24	39.805	1174.89	44.1475	1.5993	0.1959	RS
2	18	20	1171.7	44.0588	2.5437	0.2788	NM
	18	20	1171.7	44.0588	2.5437	0.2788	DE
	18	20	1171.7	44.0588	2.5437	0.2788	SA
	18	20	1171.7	44.0588	2.5437	0.2788	RS

Conclusion

This study enabled us to maximize the generator torque of the wind turbine using different regression models and stochastic optimization methods. The most suitable regression model for blade tip deflection (BTD) and tower top deflection (TTD) is defined as the second-order polynomial rational regression model, which was found with R^2 calculations and boundedness check. The most suitable regression model for generator speed (GS) and generator torque (GT) is found as the third-order polynomial rational regression model based on R^2 calculations and boundedness check. According to the optimization results, the maximum value of generator torque is 44.6829 kNm (see Table 6.8).

References

[1] Rajakumar, S. and Duarairaj, R. 2012. Optimization of wind turbine power coefficient parameters using hybrid technique. J. Inst. Eng. India Ser. 93(2): 141–149.

[2] Alpman, E. 2014. Effect of selection of design parameters on the optimization of a horizontal axis wind turbine via genetic algorithm. J. Phys.: Conf. Ser. 524 012044.

[3] Jin, X., Li, L., Ju, W., Zhang, Z. and Yang, X. 2016. Multibody modeling of varying complexity for dynamic analysis of large-scale wind turbines. Renew. Energy 90: 336–351.

[4] Jeong, J., Park, K., Jun, S., Song, K. and Ho Lee, D. 2012. Design optimization of a wind turbine blade to reduce the fluctuating unsteady aerodynamic load in turbulent wind. Journal of Mechanical Science and Technology 26(3): 827–838.

[5] Fuglsang, P. and Madsen, A. 1999. Optimization method for wind turbine rotors. Journal of Wind Engineering and Industrial Aerodynamics 80(1-2): 191–206.

[6] Benini, E. and Toffolo, A. 2002. Optimal design of horizontal-axis wind turbines using blade-element theory and evolutionary computation. Journal of Solar Energy Engineering 124(4): 357–363.

[7] Jureczko, M., Pawlak, M. and Mezyk, A. 2005. Optimization of wind turbine blades. Journal of Materials Processing Technology 167(2-3): 463–471.

[8] Casás, V., Peña, V. and Duro, R. 2006. Automatic design and optimization of wind turbine blades. Proc. of the International Conference on Computational Intelligence for Modeling Control and Automation and International Conference on Intelligent Agents Web Technologies and International Commerce (CIMCA'06), Sydney, Australia.

[9] Chaudhary, M. and Roy, A. 2015. Design & optimization of a small wind turbine blade for operation at low wind speed. World Journal of Engineering 12(1): 83–94.

[10] Maki, K., Sbragio, R. and Vlahopoulos, N. 2012. System design of a wind turbine using a multi-level optimization approach. Renew. Energy 43: 101–110.

[11] Sargolzaei, J. and Kianifar, A. 2009. Modeling and simulation of wind turbine Savonius rotors using artificial neural networks for estimation of the power ratio and torque. Simul. Model. Pract. Theory 17: 1290–1298.

[12] Abdel-Aal, R., Elhadidy, M. and Shaahid, S. 2009. Modeling and forecasting the mean hourly wind speed time series using GMDH-based abductive networks. Renew. Energy 34: 1448–1461.

[13] Monfared, M., Rastegar, H. and Kojabadi, M. 2009. A new strategy for wind speed forecasting using artificial intelligent methods. Renew. Energy 34: 845–848.

[14] Bharanikumar, R., Yazhini, A. and Kumar, A. 2010. Modeling and simulation of wind turbine driven permanent magnet generator with new MPPT algorithm. Asian Power Electron. J. 4(2): 52–58.

[15] Rao, V. and Savsani, J. 2012. Mechanical Design Optimization Using Advanced Optimization Techniques. Springer, London.

[16] Ozturk, S., Aydın, L. and Celik, E. 2018. A compherensive study on slicing process optimization of silicon ingot for photovoltaic applications. Solar Energy 161: 109–124.

CHAPTER 7

Development of Optimum Hydrogen Storage Vessels

Harun Sayi,[1] Ozan Ayakdaş,[2] Levent Aydin[3] and H Seçil Artem[1,]*

Introduction

Hydrogen has been rising as an alternative energy resource day by day, due to its bearing the highest energy density (120 MJ/kg) in terms of mass amongst all other energy sources. Additionally, hydrogen energy can solve two critical problems: (i) high carbon emissions caused by fossil fuels and (ii) energy source shortage in transportation and residence. It can be stored in various ways such as highly compressed gas, cryo-compressed liquid, or in solid storage materials. Mass of hydrogen requirement for automotive applications is approximately 4 kg for 400 km transport range.

In this case, composite pressure vessels have the attention of the automotive, marine, and aerospace industry by providing low weight and high-pressure storage vessels [1]. Increment of composite pressure vessel demands leads to developments on more reliable storage systems with higher working pressure capability. In order to achieve over the limits of the storage systems, optimization techniques with analytical and/or numerical approaches have frequently been used. Therefore, there are many studies carried out by researchers for the optimization of composite pressure vessels. The strength variables of composite hydrogen storage

---

[1] İzmir Institute of Technology, Department of Mechanical Engineering, Izmir, Turkey.
Email: harunsayi07@gmail.com
[2] Habaş Group, Aliağa.
Email: ozanayakdas@gmail.com
[3] İzmir Kâtip Çelebi University, Department of Mechanical Engineering, Izmir, Turkey.
Email: leventaydinn@gmail.com
* Corresponding author: secilartem@iyte.edu.tr

vessels can be designed and optimized by changing stacking sequences of the laminate. Statistical studies in the literature show that the application of stochastic optimization methods is dependable to solve the problems of the composites vessels [1].

Many studies in the literature demonstrated that the design of composite pressure vessels contains many problems, such as decreasing high costs and weight and increased burst strength and load capacities [2–11]. Stochastic optimization methods have been utilized in the literature to design optimum composite pressure vessels such as vessels with optimum stacking sequences, low cost, high burst pressure performance, etc. One of them is about the stacking sequences design of composite cylinders by using Genetic Algorithm (GA), which is a stochastic optimization method [12]. The similar results are obtained compared with those of experimental and numerical analysis results.

In another study, GA has been used to design a cylindrical part of the composite pressure vessel by making three-dimensional stress-strain calculations [14]. The researchers have determined the optimum winding angle on the design of the composite pressure vessels by using Classical Laminated Plate Theory (CLPT) and Tsai-Wu failure criterion [13]. As a result of the study, optimum fiber orientation angles are found as 52.1° and 54.1° for graphite/epoxy made pressure vessel (T300/N5208).

Tomasetti et al. [17] performed a stacking sequence optimization study considering Type III pressure vessels using GA by selecting objective function as the weight and the failure of the vessel. However, a single type of orientation angle has been selected among the various angle constraints. Finite Element Method (FEM) and GA have been used together to determine the minimum weight of the pressure vessel. As an uncommon optimization algorithm compared to Differential Evolution, for example, the artificial immune system (AIS) method was applied to achieve optimal weight using Tsai-Wu failure criterion as failure constraint in a study. Additionally, a single winding angle has also been used as a design variable depending on the opening radius of the composite pressure vessel having 150 MPa burst pressure [18].

There is another comparison study about the optimization of the composite pressure vessels using the Adaptive Genetic Algorithm (AGA), Simple Genetic Algorithm (SGA) and Monte Carlo (MC) search algorithms [19]. Different pressure vessel designs having a burst pressure of 164.5 MPa were proposed for different radius. In conclusion, first ply failure, and final failure criteria selection has been found to be very important to determine the optimum design procedure for Type III pressure vessels [20].

There are several studies on the optimization of Type IV pressure vessels (composite overwrapped plastic liner). Alcantar et al. [22]

performed one of these studies by dealing with the optimization of Type IV pressure vessels. One of the problems was a pressure vessel without a plastic liner. It was optimized to store 5.8 kg of hydrogen gas at 70 MPa working pressure. For other Type IV vessel designs, the aim was to optimize weight using GA and Simulated Annealing (SA). CLPT and Tsai-Wu failure criterion were utilized to design pressure vessels having a burst strength against to 70 MPa working pressure with a safety factor of 2.25. It was shown that the proposed methodology provided more efficient designs by reducing the weight up to 9.8% and 11.2%.

As a case study for hydrogen storage vessels, first-ply failure optimization of stacking sequence design of cylindrical composite overwrapped metal liner with 700-bar working pressure and 2.0 safety factor is considered. In optimization, Differential Evolution and Nelder Mead search algorithms were used to obtain desired stacking sequence design while considering Tsai-Wu, Hashin-Rotem and Maximum Stress failure theories. Three Type III pressure vessels have been considered to be having different inner volumes. Hence, the aim was to reveal the effects of a different storage volume on stacking sequence design and optimization of the cylindrical vessels according to the particular failure constraint, which is the burst pressure (≥ 140 MPa).

Mechanics of the Composite Pressure Vessels

Stress and Strain Analysis

The cylindrical part of Type III vessels is considered as a composition of the isotropic metallic liner and orthotropic composite layers. Type V vessels are also considered where the reinforcement of the vessel structure is provided by only using orthotropic composite layers. The laminated composite pressure vessel having a radius of r is subjected to the internal pressure p. The force resultants, calculated via considerations of static equilibrium [24], are

$$N_x = \frac{1}{2}\,pr \qquad\qquad N_y = pr \qquad N_{xy} = 0 \qquad\qquad (7.1)$$

where N_x, N_y and N_{xy} are axial, circumferential (hoop) and radial loading resultants, respectively.

The stress-strain analysis of Type III pressure vessels was performed by considering only membrane effects. It is assumed that strain for both materials is the same. Therefore, the extensional, stretching-coupling and flexural stiffness matrices; [A], [B], [D], was calculated by taking both structure materials; composite and metal parts, into consideration. Here, it can be seen that matrix [A] including the stiffness of the metal liner and the composite is constructed. Using matrix [A], the strains occurred due to the loadings in the vessel structure and the stresses in the liner and composite were calculated [23].

$$A_{ij} = \sum_{k=1}^{n} [(Q_{ij})]_k \, (h_k - h_{k-1}) + [Q_L] \, t_L \qquad (7.2)$$

In Equation 7.2, $[Q_L]$ is the stiffness matrix and t_L is the thickness for liner part of the vessel. In an analysis of Type V, due to be $t_L = 0$, $[A]$ matrix is calculated using Equation 7.2. The strains and $[A]^{-1}$ can be calculated using resultant loadings, N_x, N_y, N_{xy} with the relation given below as

$$\begin{bmatrix} \varepsilon_x \\ \varepsilon_y \\ \gamma_{xy} \end{bmatrix} = [A]^{-1} \begin{bmatrix} pr/2 \\ pr \\ 0 \end{bmatrix} \qquad (7.3)$$

The components of $[A]$ matrix are found using Equation 7.2 and can be represented in terms of directions of laminate as

$$[A] = \begin{bmatrix} A_{11} & A_{12} & A_{16} \\ A_{12} & A_{22} & A_{26} \\ A_{16} & A_{26} & A_{66} \end{bmatrix} \qquad (7.4)$$

After obtaining strains for laminates, using stress-strain relationships stress at the liner and each laminate can be determined with the following Equations [23].

$$\begin{bmatrix} \sigma_x \\ \sigma_y \\ \tau_{xy} \end{bmatrix} = [Q_L] \begin{bmatrix} \varepsilon_x \\ \varepsilon_y \\ \gamma_{xy} \end{bmatrix} \qquad (7.5)$$

$$\begin{bmatrix} \sigma_x \\ \sigma_y \\ \tau_{xy} \end{bmatrix} = [\bar{Q}]^{(k)} \begin{bmatrix} \varepsilon_x \\ \varepsilon_y \\ \gamma_{xy} \end{bmatrix} \qquad (7.6)$$

Failure Analysis

Failure is the inability of the structure to carry the applied load. It is caused by permanent deformation and load redistribution within the structure [25]. Failure process is the results of improper material selection, non-suitable design, improper manufacturing procedures, and variation of service condition. Theoretical background for the analysis of isotropic and anisotropic materials significantly differs from each other. Von Misses Failure Criterion is one of the most common approaches for isotropic failure analysis among Tresca and other failure criteria. If the following equation is valid, failure occurs according to the von Misses criterion.

$$(\sigma_1 - \sigma_2)^2 + (\sigma_2 - \sigma_3)^2 + (\sigma_3 - \sigma_1)^2 = 2\sigma_{yield}^2 \qquad (7.7)$$

σ_1, σ_2, σ_3, and σ_{yield} are principal stresses in main coordinate axes and yield strength (at 0.2% of strain) of the material, respectively. However, the metal part of the pressure vessel designs (liner) was assumed perfectly rigid. Therefore, the only material properties required for the analysis were stiffness and poison ratio values for this part. Since the main reinforcement of the cylindrical pressure vessel is a composite part, the failure theories dealing with anisotropic analysis must be given in detail. Tsai-Wu, Hashin-Rotem, and Maximum stress criteria have been accounted for the failure analysis.

Tsai-Wu Failure Criterion

This theory assumes that the failure of the material occurs when the equation given below is satisfied.

$$F_1\sigma_1 + F_2\sigma_2 + F_{11}\sigma_1^2 + F_{22}\sigma_2^2 + F_{66}\tau_{12}^2 + 2F_{12}\sigma_1\sigma_2 = 1 \tag{7.8}$$

where F values are the summation of specified strength. The parameters appearing here can be calculated as [26]

$$F_1 = \frac{1}{\left(\sigma_1^T\right)_{ult}} + \frac{1}{\left(\sigma_1^C\right)_{ult}} \qquad F_{11} = -\frac{1}{\left(\sigma_1^T\right)_{ult}\left(\sigma_1^C\right)_{ult}} \tag{7.9}$$

$$F_2 = \frac{1}{\left(\sigma_2^T\right)_{ult}} + \frac{1}{\left(\sigma_2^C\right)_{ult}} \qquad F_{22} = -\frac{1}{\left(\sigma_2^T\right)_{ult}\left(\sigma_2^C\right)_{ult}} \tag{7.10}$$

$$F_{12} = -\frac{1}{2}\sqrt{F_{11}F_{22}} \qquad F_{66} = \frac{1}{\left(\tau_{12}^F\right)_{ult}^2} \tag{7.11}$$

Hashin-Rotem Criterion

This theory expresses different failure patterns for composite structures which are fiber failure in tension, fiber failure in compression, matrix failure in tension, and matrix failure in compression [27].

Fibre failure in tension ($\sigma_1 > 0$)

$$\sigma_1 = (\sigma_1^T)_{ult} \tag{7.12}$$

Fibre failure in compression: ($\sigma_1 < 0$)

$$-\sigma_1 = (\sigma_1^C)_{ult} \tag{7.13}$$

Matrix failure in tension: ($\sigma_2 > 0$)

$$\left(\frac{\sigma_1}{(\sigma_2^T)_{ult}}\right)^2 + \left(\frac{\tau_{12}}{(\tau_{12})_{ult}}\right)^2 = 1 \tag{7.14}$$

Matrix failure in tension: $(\sigma_2 < 0)$

$$\left(\frac{\sigma_2}{(\sigma_2^C)_{ult}}\right)^2 + \left(\frac{\tau_{12}}{(\tau_{12})_{ult}}\right)^2 = 1 \qquad (7.15)$$

The Maximum Stress Theory

This criterion predicts failure through assumptions based on the relation of principal stress values $(\sigma_1, \sigma_{12}$ and $\tau_{12})$ and ultimate strength parameters of the materials. Tensile, compression and shear failure models expressed as [28].

$$\sigma_1 \leq (\sigma_1^T)_{ult}, \ \sigma_2 \leq (\sigma_2^T)_{ult} \qquad \text{if} \quad (\sigma_1 > 0, \ \sigma_2 > 0) \qquad (7.16)$$

$$|\sigma_1| \leq (\sigma_1^C)_{ult}, \ |\sigma_2| \leq (\sigma_2^C)_{ult} \qquad \text{if} \quad (\sigma_1 < 0, \ \sigma_2 < 0) \qquad (7.17)$$

$$|\tau_{12}| \leq (\tau_{12})_{ult} \qquad (7.18)$$

where $(\sigma_1^T)_{ult}$, $(\sigma_2^T)_{ult}$, $(\sigma_1^C)_{ult}$ and $(\sigma_2^C)_{ult}$ are ultimate strength values for an anisotropic material. $(\tau_{12})_{ult}$ is the ultimate shear strength value for a specified material. 1 stands for in fiber direction while 2 is for transverse fiber direction. Notation of T represents tensile strength direction, and C is for compression type of strength.

As an application study, the stacking sequence design of filament wound composite pressure vessel with an aluminum liner was solved using Differential Evolution and Nelder Mead search algorithms. CLPT and first-ply failure approach were incorporated to determine stress and strains. Though, three deterministic designs differing in storage volume have been optimized to perform working pressure of 700 bar and safety factor of 2.0 (failure pressure ≥ 1400 bar).

The aim was to optimize the stacking sequences of the designs to maintain the working pressure and the safety factor. Furthermore, as a result of changing storage volume, consequences of implementing different stochastic search algorithm, and failure criteria on stacking sequence design were revealed. Figure 7.1 shows the specific parameters and constant values such that inner radius (r), cylindrical coordinates, internal pressure (P) and angle of fibers. Geometrical properties are given in Table 7.1.

Hashin-Rotem, Tsai-Wu, and Maximum Stress failure theories were incorporated to analyze the first-ply failure of the vessels. Differential Evolution (DE) and Nelder Mead (NM) were the random search algorithms in the optimization process.

The pressure vessels must perform reliably in given conditions. Therefore, advanced composite systems were utilized. Carbon fiber reinforced epoxy polymer made the composite part of the vessels while aluminum 6061 T6 formed the liner part. Mechanical properties of the vessel structures are given in Table 7.2.

The variable parameters in the optimization process were set as the amount of layers and the angles of fiber (θ). One of the conditions to use CLPT in analyzing stress-strain behavior was the requirement of the thin-walled pressure vessel. Therefore, the wall thickness of each design must satisfy the thin-walled vessel theory; the ratio of total wall thickness to inner radius must not exceed 1/10. As a result, there was a limitation

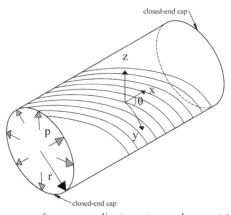

Figure 7.1 Inside pressure, reference coordinate system and geometrical parameters of the vessel designs [24].

Table 7.1 Geometrical properties of the design problems.

Designs	Volume	Inner Radius	The Thickness of a Ply	Thickness of Liner	Length
Design I	5	56.42			
Design II	10	79.79	0.127	3	500
Design III	20	112.94			

* Unit of the volume is in lt (Liter).
* Unit of length is in mm (millimeter).

Table 7.2 Elastic properties of materials used in case study designs [23].

Material	Mechanical Properties									
	E_1	E_2	G_{12}	v_{12}	v_{21}	$\sigma^{ult}_{1,T}$	$\sigma^{ult}_{1,C}$	$\sigma^{ult}_{2,T}$	$\sigma^{ult}_{2,C}$	τ^{ult}_{12}
T700/Epoxy	181	10.3	5.86	0.28	0.016	2150	2150	298	298	778
6061 Al	70	70	26.92	0.3			−			

* Units of E_1, E_2, G_{12} are in GPa.
* Unit for strength parameters is MPa.

in the number of layers for each optimized design. It must not exceed 20 layers for Design I, 39 layers for Design II, and 65 layers for Design III. The optimization results of corresponding cylindrical vessel designs were shown in Table 7.3.

Find: $\{\theta_p, n\}$, $\theta_p \in [\theta_1, -\theta_1, \ldots, \theta_n, -\theta_n]_s$, $\theta_p \in$ Integers, where number of layers for;

Design I ≤ 20,

Design II ≤ 39,

Design III ≤ 65,

Minimize: Tsai-Wu, Hashin-Rotem and Maximum Stress failure index

Maximize: Failure pressure of the vessels

Constraints: $FI_{\text{Tsai-Wu}} \leq 1$, $FI_{\text{Hashin-Rotem}} \leq 1$ *where FI:Failure Index*

Maximum stress ≤ 1 (for each ply)

$-90 \leq \theta_p \leq 90$

Symmetry

Methods: Differential Evolution and Nelder-Mead

Iteration Number: 500

In the first design, DE and NM have obtained significantly different solutions. Optimization performed by DE required only two composite plies to achieve design targets. In the opposite direction, NM has required 12 layers more to comprise the constraints. Hence, the efficiency of NM in this design was below compared to DE. This case was also valid while comparing CPU speeds. In conclusion, the time and cost-efficient method in optimizing first problem design was using DE with the incorporation of the Tsai-Wu criterion.

The difference between the optimum stacking sequence obtained for algorithms was marginal in design II. NM was quicker to obtain results. Fibers were located at around 90° angle. Both algorithms provided the minimum number of plies when failure constraint was Tsai-Wu. It should be said that the optimization process with using NM and Tsai-Wu has been the most efficient when cost and time were taken into consideration.

In the third design, the required number of plies and the variety of optimum angles were increased due to the expansion of storage volume. The main difference between algorithms and selected failure criteria was the completion time of observing optimum results. When all variables and completion time of the process were taken into consideration, the most efficient way to obtain design III was using NM algorithm and Hashin Rotem failure criterion.

Table 7.3 Results of stacking sequence optimization for Design I, Design II and Design III.

Design	Failure Constraint	Stacking Sequence (DE)	# of Ply (DE)	Stacking Sequence (NM)	# of Ply (NM)	CPU Time (DE)	CPU Time (NM)
I	Tsai-Wu	$[21]_S$	2	$[\pm28/\mp43_2]_S$	12	3	16.5
	Hashin-Rotem	$[\mp28]_S$	4	$[\mp2/\pm33/\pm34/-39]_S$	14	4.1	19.4
	Maximum Stress	$[\mp27]_S$	4	$[\mp2/\mp35/\mp10/\pm35]_S$	16	5.2	26.2
II	Tsai-Wu	$[\mp90_6]_S$	24	$[\pm90/\mp90/\pm90_2/\mp90/\pm90]_S$	24	64.7	48.6
	Hashin-Rotem	$[\pm90/\mp90/\pm90_5]_S$	28	$[\mp90_2/\pm89_2/\mp89/\pm90_2]_S$	28	75.9	62.7
	Maximum Stress	$[\pm90/\mp90/\pm90_2/\mp90_2/\mp90/\pm90]_S$	28	$[\mp90/\mp89/\pm87/\pm88/\mp86/\pm89/\pm88]_S$	28	83.5	72.6
III	Tsai-Wu	$[\pm90/\pm72/\mp72/\pm72/\mp90/\mp73/\mp72_2/\pm72/\mp72/-90]_S$	42	$[\mp73_2/\pm73/\mp83/\pm73_2/\pm75/\mp81/\mp90/\mp73/-88]_S$	42	255.1	201.8
	Hashin-Rotem	$[\mp90_8/\pm90_2/\mp90]_S$	44	$[\mp90_2/\pm90/\mp90/\pm90_3/\mp90_3/\pm90]_S$	44	178.9	163.3
	Maximum Stress	$[\mp90/\pm90/\mp90/\pm90/\mp90_6/\pm90]_S$	44	$[\mp90/\mp89/\pm89/\mp90/\pm89/\mp90/\pm88/\mp89_4]_S$	44	194.8	189.2

* Unit of time is in second.

Conclusion

This chapter has presented a review of high-pressure storage vessels. Besides that, it included a case study about the design and analysis of stacking sequences for composite pressure vessels by optimization algorithms. The vessels were subjected to internal pressure causing stress and strain through the hoop, axial, and radial directions. The optimization procedure was set to provide the working pressure of 700 bar and a safety factor of 2. Application of the optimization methods to composite pressure vessel designs enables us to tailor mechanical properties of the structure efficiently.

Eighteen optimization problems have been performed using DE and NM algorithms incorporating Classical Lamination Plate Theory and failure theories. Failure criteria have been taken as the objective function to be minimized. The total number of plies and angle of fibers were the variable parameters in the optimization setup. Fiber angle orientation has been taken as angle-ply formation $\left[\pm\theta_1 / ... / \pm\theta_n \right]$ and symmetry of stacking sequence have been maintained for design problems considered. Three Type III cylindrical hydrogen storage vessel designs have been considered. Material properties, the thickness of a ply and thickness of liner were identical for each design. Composite part of the vessel was made by carbon fiber/epoxy while the liner was selected from aluminum 6061 T6. The storage vessels with 5, 10 and 20 liters have been considered for the specified problems Design I, Design II and Design III, respectively. There was no violation in optimized vessels concerning thin-walled cylindrical vessel theory. A computer code based on CLT employing failure theories and optimization algorithms has been developed by using Wolfram MATHEMATICA Software. Three different failure approaches have been considered, which are an interactive failure theory; Tsai-Wu, a partially interactive criterion; Hashin-Rotem and Maximum Stress as the non-interactive criterion. Another objective of the study has been to reveal the effect of volume differences on the fiber orientation angle and the number of plies in cylindrical composite pressure vessels.

According to the results of the optimization process, the efficient design concerning the required number of layers and less variety in fiber orientation angles for manufacturing purposes, one might prefer to set optimization using Differential Evolution. However, Nelder Mead determined solutions in less time. In the failure criterion perspective, optimizations performed using Hashin-Rotem failure theory as constraint were obtained in less time on average.

References

[1] Cohen, D. 1997. Influence of filament winding parameters on composite vessel quality and strength. Compos. Part A 28(A): 1035–1047.

[2] Chapelle, D. and Perreux, D. 2006. Optimal design of a type 3 hydrogen vessel: Part I Analytic modelling of the cylindrical section. Int. J. Hydrogen Energy 31(5): 627e38.

[3] Hu, J., Chen, J., Sundararaman, S., Chandrashekhara, K. and Chernicoff, W. 2008. Analysis of composite hydrogen storage cylinders subjected to localized flame impingements. Int. J. Hydrogen Energy 33(11): 2738e46.

[4] Onder, A., Sayman, O., Dogan, T. and Tarakcioglu, N. 2009. Burst failure load of composite pressure vessels. Compos. Struct. 89(1): 159e66.

[5] Xu, P., Zheng, J. and Liu, P. 2009. Finite element analysis of burst pressure of composite hydrogen storage vessels. Mater. Des. 30(7): 2295e301.

[6] Hu, J. and Chandrashekhara, K. 2009. Fracture analysis of hydrogen storage composite cylinders with liner crack accounting for autofrettage effect. Int. J. Hydrogen Energy 34(8): 3425e35.

[7] Camara, S., Bunsell, A.R., Thionnet, A. and Allen, D.H. 2011. Determination of lifetime probabilities of carbon fiber composite plates and pressure vessels for hydrogen storage. Int. J. Hydrogen Energy 36(10): 6031e8.

[8] Son, D.S. and Chang, S.H. 2012. Evaluation of modeling techniques for a Type III hydrogen pressure vessel (70 MPa) made of an aluminum liner and a thick Carbon/Epoxy composite for fuel cell vehicles. Int. J. Hydrogen Energy 37(3): 2353e69.

[9] Liu, P., Chu, J., Hou, S. and Zheng, J. 2012. Micromechanical damage modeling and multiscale progressive failure analysis of composite pressure vessel. Comput. Mater. Sci. 60: 137e48.

[10] Leh, D., Saffre, P., Francescato, P. and Arrieux, R. 2013. Multi-sequence dome lay-up simulations for hydrogen hyper-bar composite pressure vessels. Compos. Part A Appl. Sci. Manuf. 52: 106e17.

[11] Liu, P.F., Chu, J.K., Hou, S.J., Xu, P. and Zheng, J.Y. 2012. Numerical simulation and optimal design for composite high-pressure hydrogen storage vessel: a review. Renew. Sustain. Energy Rev. 16(4): 1817e27.

[12] Messager, T., Pyrz, M., Gineste, B. and Chauchot, P. 2002. Optimal laminations of thin underwater composite cylindrical vessels. Compos. Struct. 58: 529–37.

[13] Parnas, L. and Katirci, N. 2002. Design of fiber-reinforced composite pressure vessels under various loading conditions. Compos. Struct. 58: 83–95.

[14] Tabakov, P.Y. 2001. Multi-dimensional design optimization of laminated structures using an improved genetic algorithm. Compos. Struct. 54: 349–54.

[15] Richard, F. and Perreux, D. 2000. A reliability method for optimization of [+h/_h] n fiber reinforced composite pipes. Reliab. Eng. Syst. Saf. 68: 53–9.

[16] Kim, C.U., Hong, C.S., Kim, C.G. and Kim, J.Y. 2005. Optimal design of filament wound type 3 vessels under internal pressure using a modified genetic algorithm. Compos. Struct. 71(1): 16e25.

[17] Tomassetti, G., Barboni, R. and Benedetti, M. 2005. Optimisation methodology for cryovessels. Comput. Struct. 83(28): 2293e305.

[18] Liu, P., Xu, P. and Zheng, J. 2009. Artificial immune system for optimal design of composite hydrogen storage vessel. Comput. Mater. Sci. 47(1): 261e7.

[19] Xu, P., Zheng, J., Chen, H. and Liu, P. 2010. Optimal design of high pressure hydrogen storage vessel using an adaptive genetic algorithm. Int. J. Hydrogen Energy 35(7): 2840e6.

[20] Francescato, P., Gillet, A., Leh, D. and Saffré, P. 2012. Comparison of optimal design methods for type 3 high-pressure storage vessels. Compos. Struct. 94: 2087–2096.

[21] Roh, H., Hua, T. and Ahluwalia, R. 2013. Optimization of carbon fiber usage in type 4 hydrogen storage vessels for fuel cell automobiles. Int. J. Hydrogen Energy 38(29): 12795e802.

[22] Alcantar, V., Aceves, S.M., Ledesma, E., Ledesma, S. and Aguilera, E. 2017. Optimization of type 4 composite pressure vessels using genetic algorithms and simulated annealing. Int. J. Hydrogen Energy 0360–3199.

[23] Alcantar, V., Ledesma, S., Aceves, S.M., Ledesma, E. and Saldana, A. 2017. Optimization of type III pressure vessels using genetic algorithm and simulated annealing. Int. J. Hydrogen Energy 42: 20125–20132.

[24] Pelletier, J.L. and Vel, S.S. 2006. Multi-objective optimization of fiber-reinforced composite laminates for strength, stiffness and minimal mass. Comput. Struct. 84: 2065–2080, doi: 10.1016/j.compstruc.2006.06.001.

[25] Hyer, M.W. and White, S.R. 1998. Stress Analysis of Fiber-Reinforced Composite Materials. WCB McGraw-Hill, ISBN: 0070167001.

[26] Aydin, L., Artem, H.S., Oterkus, E., Gundogdu, O. and Akbulut, H. 2017. Mechanics of fiber composites, Fiber Technology for Fiber Reinforced Composite. Woodhead Publishing Press, Chapter 2, 5-48, ISBN: 978-0-08-101871-2.

[27] Hashin, Z. and Rotem, A. 1973. A fatigue failure criterion for fibre reinforced materials. J. Compos. Mater. v.7: 448–464.

[28] Vasiliev, V.V. and Morozov E.V. 2007. Advanced Mechanics of Composite Materials. ISBN: 978-0-08-045372-9, Elsevier.

CHAPTER 8

Optimization of Surface Roughness in Slicing Process of Silicon Ingots

Savas Ozturk,[1,]* *Nilay Kucukdogan,*[2]
Levent Aydin[3] and *Ayşe Seyhan*[4,5]

Introduction

In today's world, one of the most critical industrial targets of countries is energy production. Although large-scale thermal, nuclear, and hydropower plants meet most of the market demand in energy production, the increasing need for energy necessitates the use of alternative energy sources. In this sense, solar energy, which stands out amongst others as a source of clean energy, is exciting especially with its increasing potential. As an application of solar energy, solar cells can be produced in many different materials and different ways. Gallium arsenide and multi-junction cells are among the most exciting options with their high yields, but because

[1] Department of Metallurgical and Materials Engineering, Manisa Celal Bayar University, Manisa, 45140, Turkey.
[2] The Graduate School of Natural and Applied Sciences, Izmir Katip Celebi University, 35620, Izmir, Turkey.
Email: kucukdogan.nilay@gmail.com
[3] Department of Mechanical Engineering, Izmir Katip Celebi University, Izmir, 35620, Turkey.
Email: leventaydinn@gmail.com
[4] Department of Physics, Nigde Omer Halisdemir University, Nigde, 051240, Turkey.
Email: aseyhan@ohu.edu.tr
[5] Labotecha Technology Consulting Company, Nigde Omer Halisdemir University, Technopark A.S., Center Campus, No:10, Nigde, 051240, Turkey.
* Corresponding author: savas.ozturk@cbu.edu.tr

of their high cost, they are more suitable for high-tech applications such as spacecraft and, unmanned aerial vehicles. The most widely used solar cells are silicon-based ones with single or multi-crystalline types. These solar cells are fabricated on the prepared surfaces of thin wafers, which are produced by slicing Si ingots using different methods. In terms of cell yield, the surface quality of the Si wafer is crucial, and surface roughness and micro-crack density are the most evaluated features for measuring surface quality. Slicing is generally done with multi or single wire slicing systems. Parameters such as slicing speed, wire feed speed, wire tension, size, and density of abrasives are critical production parameters for slicing. Slicing wafers with minimum crack density and surface roughness is very critical for reducing the cost of solar cell production and improving cell performance.

In this section, the effects of process parameters (slicing speed and wire feed rate) on surface roughness are studied in the slicing process of single-crystalline Si ingot into thin Si wafers using single embedded diamond wire slicing system. For optimization and functional determination of wafer surface roughness (R_a), a nonlinear mathematical model based on Neuro-regression approach was created by using Wolfram Mathematica v.11, and then the surface roughness was correlated with slicing parameters. The solution to the slicing problem being studied is arranged in three stages. Firstly, a mathematical model has been developed to estimate the surface roughness of the slicing surface during the abrasive slicing of silicon ingot with diamond wire saw. Secondly, the predictability of Si ingot slicing model was investigated. Finally, DE, NM, SA, and RS algorithms were used in the optimization section for minimizing surface roughness in the slicing process.

Literature Survey

In the world, energy can be produced by using different energy sources according to the underground wealth and natural conditions [1]. Accessibility, reliability, and sustainability of energy resources are essential in determining energy policies. If a simple comparison is made between these sources in terms of affordability, the prices of petroleum and derivatives are increasing, whereas solar and wind power is decreasing with each passing year. Reliability of energy sources is another essential issue and political turmoil, especially in countries with high oil reserves, and causes fluctuations in the oil supply. Because the emission of global energy-based carbon dioxide increased by 1.6% in 2017, the harm of carbon-containing fuels to global warming and human health is very worrying and not sustainable in the long term [2]. Solar energy systems have an increased potential for use as a critical non-emission energy source that can be easily purchased due to their reduced costs [3].

The World Energy Outlook (WEO), which is referred to as the "New Policies Scenario", states that the number of people migrating to urban areas in developing countries will be 1.7 billion by 2040, and this displacement will increase the energy requirements by at least 25%. It also states that renewable energy sources will be of great importance in meeting this energy need. Also, the competitiveness of solar photovoltaics (PV) is expected to increase rapidly, and its installed capacity will exceed wind before 2025, hydroelectric energy around 2030 and coal before 2040 [2].

Solar PV systems are produced as a wide variety of cells with the difference in efficiency values and costs. These include amorphous silicon (a-Si), cadmium telluride (CdTe), copper indium gallium selenide (CI(G)S), dye-sensitized (DSSC), organic (OPV), perovskite, and quantum dot solar cells, which have attracted attention in recent years with its increasing efficiency values. When yield values are compared, multi-junction Si solar cell stands out with yield values up to 50% [4]. However, since its availability worldwide is directly dependent on cost, about 90% of installed PV systems are mono or multi-crystalline Si solar cells [5].

Approximately 95% of the solar panels installed in the world consist of Si-based cells. Approximately 56% of these cells are multi-Si, 36% of them are mono-Si, and the rests are amorphous silicon and its derivatives. The production of crystalline Si solar cells involves many painful and expensive process steps starting from quartz sand. Purification of quartz at high temperatures and subsequent bulk Si (Si ingot) production is the first step. The second step includes the process for slicing of Si ingot. In crystalline solar cells, cell thickness directly affects the yield. For this purpose, the cells are produced from Si wafers with a thickness range of 150–250 μm. After slicing, lapping and texturing there are the other steps of preparing Si wafer for solar cells fabrication. These are called wafer preparation processes, and then the process of cell production is performed [6].

Two types of slicing systems in use are conventional inner diameter (ID) saws and wire saws. ID saw systems ceased to be used in the 1990s due to low slicing efficiency, increased surface damage, and high kerf loss [5]. Since then, wire systems have dominated wafer production. Wire systems are classified into two main categories as single and multiple slicing systems [6]. The schematic view of the wire saw is shown in Figure 8.1a. In multi-wire systems (MWS), the abrasive is fed onto the wires as a solution (see Figure 8.1b), whereas in single-wire (or diamond wire) systems (DWS) the abrasive is embedded in the wire (Figure 8.1c). Most of the wafer production is carried out in multi-wire systems, and the obtained wafer surfaces are more homogeneous since the abrasives are controlled in terms of size and density distributions [7]. In some

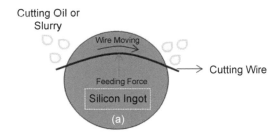

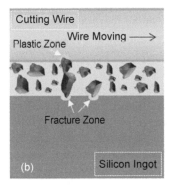

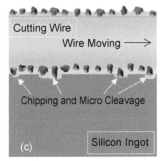

Figure 8.1 (a) Schematic views of wire sawing, (b) cutting zone of multi-wire slurry saw, and (c) cutting zone of diamond-wire saw.

academic publications [5, 7, 8], it is said that surfaces obtained in single-wire systems have less roughness and damage than those of multi-wire systems. The position of the wire in use changes the characteristics of the cut surface because the abrasives in diamond-embedded single-wire systems are deformed when the wire is in use. In both the MWS and DWS systems, the slicing wire is fed back after feeding a certain distance, called "Reciprocating motion". This behavior causes a wavy surface to form on the slicing surface [7]. The reciprocating motion of diamond wire during the DWS process is shown in Figure 8.2. In this figure, t_{cy} is the cycle time, and t_{cf} and t_{cb} are the times for the constants to feed-forward and backward, respectively.

The efficiency of the Si solar cell is directly related to the perfection of the surface [3]. Surface defects during the preparation of wafer reduce the cell's efficiency and service life. To overcome these drawbacks, they have performed lapping and polishing operations after slicing. Obtaining less rough and less damaged surfaces in the slicing process reduces lapping and polishing times, thus reducing the cost of wafer preparation [3, 9]. Many parameters of single wire slicing systems affect the surface roughness values of the produced wafers. The feeding speed and tension of the wire, the slicing speed of the wafer, the size, quantity, and shape

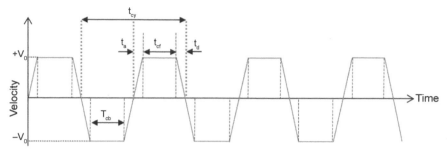

Figure 8.2 The reciprocating motion of diamond wire during the slicing process.

of the abrasive embedded in the wire, the viscosity, and type of the used coolant are among the most critical parameters [3].

There are many academic studies about Si slicing process with a wire saw. In these studies, polycrystalline [10] and monocrystalline [5, 7, 8, 11] Si ingots were used as slicing material, and also both DWS and MWS were used as slicing systems. Wire speed, feed rate, and the properties of abrasives and wires are the most commonly studied slicing parameters. In these studies, surface roughness values were examined, and the characteristics of kerf loss and defect formation were investigated [5].

Surface roughness, wear rate and amount of wear are subjects that are investigated in optimization studies in machining applications. In the literature, the abrasion behavior of metallic materials has been studied a lot, while there are very few studies on the cutting of Si. Ozturk et al. 2018 [5] reported some outstanding publications in the literature for machining metallic materials based on optimization algorithms. In these studies, steel, titanium, aluminum, copper and nickel alloys were taken into consideration as machining materials; regression analysis and response surface methodology were selected to generate target functions; Genetic Algorithm, Gauss-Newton Algorithm, Particle Swarm, Sheep Flock, Ant Colony, Artificial Bee Colony, Biogeography-Based, and Simulated Annealing algorithms were performed in the optimization part of the studies. In all these studies, in general, a single objective function was used, while Ozturk et al. 2018 [5] created eight different forms of regression and operated them with four different optimization algorithms for the minimization of surface roughness in the Si slicing process. In this study, they stated that it is not correct to describe every engineering process with only polynomial models, and therefore, the different models that define them may describe the engineering phenomena better. Besides, the four algorithms run with each model created in the optimization studies increased the reliability of the process and prevented the natural scattering of stochastic algorithms [5].

Objectives and Motivation of the Study

The slicing operation is a vital process for turning a-Si ingot or block into a wafer, and there is almost a fifty percent loss of material depending on the slicing system and wire type. In the case of the slicing of brittle materials, surface damage such as plastic rupture zones and wear marks occur in the slicing zone. Si, which is a brittle material, is the most commonly used substrate material, especially in solar cell production. The presence of damage to the slicing surface results in surface recombination, which degrades the performance of the solar cell. To avoid this problem, the damages on the surface of the Si substrate and under the surface it must be removed before the device production. Slicing and subsequent surface damage removals are the significant causes of material loss in Si. If the damage, roughness, or both values on the surface of the material are considerable during the slicing process, both the amount of the lost material and the processing time for eliminating them would be high. Reducing costs in solar cell production increases the usage of solar energy as an alternative energy source. Therefore, if damage to the Si surface by slicing could be minimized, the cost to eliminate this damage will be reduced. A good understanding of the slicing process and a good optimization of the slicing parameters will minimize the amount of damage to the slicing surface.

The slicing, lapping, polishing, and texturing processes are all evaluated in the calculation of total time and cost in the preparation of Si wafer for solar cells. Problems that occur in one stage are eliminated in the next one. For example, high surface roughness during slicing increases the lapping time. The most important factors that increase material loss and the cost for wafers cut with DWS system is the breakage of the wire during the slicing process (i.e., the deformation of the slicing equipment and surface). For this purpose, it is essential to optimize the slicing process parameters. This operation increases the life of the slicing parts and minimizes the time and cost of wafer surface preparation.

The present study considers obtaining the optimal process parameters of slicing of monocrystalline Si ingot in wafer form. In the modeling, the experimental data were taken from Ozturk et al. 2018 [5], in which the effects of slicing parameters on surface roughness values were investigated.

Definition of the Problem

The quality of Si wafers (within a certain amount of dopant) used in solar cells and electronic devices is directly related to the morphological properties of the surface. Between the Si wafer production stages, most of the surface damage occurs in slicing with a wire saw. In this process, abrasive particles on the cutting surfaces cause abrasion damage. Furthermore, wave-like wear marks are formed on the cutting surface during the back and forth of the cutting wire. The surface roughness value R_a can be defined as the arithmetic mean of the absolute values of the profile deviations from the mean line, which is determined by surface roughness measurement techniques. The mathematical expression of R_a can be expressed as in Equation 8.1.

$$R_a = \frac{1}{n} \sum_{i=1}^{n} |y_i| \tag{8.1}$$

In this study, experimental data of the wire speed (X_{ws}) and feed rate (X_{fr}) in the DWS system from Ozturk et al. 2018 [5] (see Table 8.2) was used. It should be noted that in their experimental studies, they replaced the cutting wires with every ten cuts, and the arithmetic means of the cut wafers were evaluated. In the present optimization study, objective function, constraints, and design variables of the problems are listed in Table 8.1.

Table 8.1 Objective functions, constraints and design variables of the optimization problems.

Problems	Objectives	Constraints	Design Variables
1	Surface Roughness R_a minimization	$(2.5 < X_{ws} < 4.5)$ $(0.5 < X_{fr} < 1.0)$	X_{ws} X_{fr}
2	Surface Roughness R_a minimization	$X_{ws} \in \{2.5, 3, 3.5, 4, 4.5\}$ $X_{fr} \in \{0.5, 0.75, 1.0\}$	X_{ws} X_{fr}

Regression Analysis

The effect of experimental parameters of wire speed and feed rate on surface roughness (R_a) value in cutting the wafer form of Si ingot with DWS system was used in the preparation of second-order nonlinear polynomial objective function. Regression analysis and "Artificial Neural Network" (ANN) methods were used together in "hybrid" form, which is also called neuro-regression to obtain objective function. For this purpose, in order to model R_a as the output of the system, 12 experimental sets were used to form objective functions by calculating R^2 training, and the remaining three experimental sets were used for testing with R^2 testing calculation as in ANN systematics.

The general form of regression models (Equation 8.2) prepared for training data is as follows:

$$Y = \frac{a_0 + a_1 X_{ws} + a_2 X_{fr} + a_3 X_{fr} X_{ws} + a_4 X_{fr}^2 + a_5 X_{ws}^2}{a_6 + a_7 X_{ws} + a_8 X_{fr} + a_9 X_{fr} X_{ws} + a_{10} X_{fr}^2 + a_{11} X_{ws}^2} \tag{8.2}$$

Table 8.2 Slicing parameters and an average of measured surface roughness values in the prepared Si wafers by the DWS device [5].

		Slicing Parameters		Measured Surface Roughness Values
	Experimental Run Numbers	Wire Speed (m/s)	Feed Rate (mm/min)	R_a (μm)
Training	1	2.5	0.5	3.26
	2	2.5	0.75	2.03
	3	2.5	1.0	2.07
	4	3.0	0.5	1.14
	5	3.0	0.75	1.43
	6	3.0	1.0	2.21
	7	3.5	0.5	1.15
	8	3.5	0.75	1.49
	11	4.0	0.75	1.78
	12	4.0	1.0	2.30
	14	4.5	0.75	1.83
	15	4.5	1.0	2.90
Testing	9	3.5	1.0	2.17
	10	4.0	0.5	1.08
	13	4.5	0.5	1.25

where X_{ws} and X_{fr} represent the parameters wire speed and feed rate, respectively.

$$Y = \frac{-2.594 - 23.871\ X_{fr} + 18.084\ X_{fr}^2 + 5.726\ X_{ws} + 0.189\ X_{fr}X_i - 0.764\ X_{ws}^2}{-3.761 - 10.071\ X_{fr} + 8.998\ X_{fr}^2 + 4.131\ X_{ws} - 0.696\ X_{fr}X_{ws} - 0.477\ X_{ws}^2}$$

(8.3)

The R^2 training value, which prepared with the training data for the model given in Equation 8.3 was found to be 0.996. When the same data was run with the model prepared by Ozturk et al. 2018 [5], R^2 was calculated as 0.988. Parameters were calculated with the model prepared for testing the data. The relationship between these values and experimental values was examined statistically, and R^2 testing value was found to be 0.78. Having a high R^2 value during the testing stage shows that this model is suitable and describe this engineering phenomenon very well.

Optimization of Si Ingot Slicing Parameters

Four different optimization algorithms studied the problem of minimizing surface roughness in the cutting process. Optimization studies were carried out in two different problem solutions according to the parameter limitations of the device used in the experimental studies. The first problem is designed without the limitations of the device parameters, and the second problem is constructed with device constraints. Since there is no parameter value limitation in the first problem, the system can find the minimum surface roughness at the levels of 10^{-9} µm (see Table 8.3). In the second problem solving, predicted values close to the experimental R_a values were obtained (see Table 8.4).

The results from the present study, together with the optimization results in the reference article are given in Table 8.4. The limit conditions for the parameters were determined according to the experimental parameter limits in the reference article. The optimization algorithms

Table 8.3 Optimum slicing parameters and R_a values of current studies for the first problem.

Optimization Algorithms	R_a Values (µm)	Feed Rate (mm/min)	Wire Speed (m/s)
DE	1.561×10^{-9}	0.500008	4.98798
NM	6.282×10^{-9}	0.61588	4.79359
RS	1.097×10^{-9}	0.53266	4.90823
SA	1.426×10^{-9}	0.604772	4.80019

Table 8.4 Optimum slicing parameters and R_a values of reference [5] and current studies for the second problem.

Optimization Algorithms	Study	R_a (µm)	Feed Rate (mm/min)	Wire Speed (m/s)
DE	Current	0.082	0.60	2.84
	Reference	0.105	0.61	2.71
NM	Current	0.808	0.60	4.50
	Reference	0.137	0.55	2.65
RS	Current	0.181	0.60	2.85
	Reference	0.845	0.60	2.85
SA	Current	0.340	0.60	2.87
	Reference	0.112	0.58	2.87

were run according to the constraint that the minimum roughness cannot be less than 0.01 µm. This value is appropriate for cutting with diamond wire. Among the algorithms, the lowest R_a value was obtained by the DE algorithm, followed by RS, SA, and NM, respectively. The predicted value of NM algorithm is very close to the experimentally lowest value of 1.08 µm, and it is much higher than the result obtained in the reference article.

The parameters of the lowest R_a value of both the optimization problems were examined, the optimum feed rate of 0.6 mm/min was obtained as the best result in both the reference study and the current study. The wire speed was found to be about 2.8 m/s, but according to experimental data, it was interesting to obtain the lowest surface roughness value at 4 m/s. When the results of four different optimization algorithms are examined, it is seen that the NM method gives worse results than the other three stochastic methods.

Conclusion

In this study, stochastic optimization algorithms of Differential Evaluation, Nelder-Mead, Random Search, and Simulated Annealing were systematically studied in a standard cutting process. Si wafer, which is used in electronic devices, and has to be produced with almost perfect surface properties and in order to achieve this, the cutting process which is the first step of production must be performed correctly. For this purpose, the effects of cutting process parameters on surface roughness were investigated.

The experimental data examined, showed that the minimum R_a value as 1.08 µm was obtained with the cutting speed 0.5 mm/min and the wire speed with 4 m/s parameters. In the modeling section, the second-order polynomial form with 0.998 R^2 value as the best match with training data was prepared by nonlinear regression analysis. It should be noted that the cutting process, which is a critical engineering application, is defined in a rational form, unlike standard polynomial regression forms. In the optimization section, surface roughness value was obtained at levels of 10^{-8} µm in first problem solving, but these results are not to be reached in real engineering applications. The closest optimization R_a result as 0.808 µm to the lowest experimental value was obtained by using the NM algorithm with cutting speed 0.6 mm/min and wire-speed 4.5 m/s for second problem-solving. Besides, the lowest surface roughness values were obtained with the DE algorithm both in the reference article and in the present study.

References

[1] Kayabasi, E., Ozturk, S., Kücükdogan, N. and Savran, M. 2018. Recent developments in energy and solar energy. pp. 1883–1893. *In*: Arapgirlioglu, H., Atik, A., Hiziroglu, S., Elliot, R.L. and Atik, D. [eds.]. The Most Recent Studies in Science and Art. Gece Publishing, Ankara, Turkey.

[2] International Energy Agency (IEA). 2018. World Energy Outlook 2018. Retrieved from https://www.iea.org/weo2018/.

[3] Ozturk, S., Aydin, L., Kucukdogan, N. and Celik, E. 2018. Optimization of lapping processes of silicon wafer for photovoltaic applications. Sol. Energy 164: 1–11.

[4] Kirchartz, T. and Uwe, R. 2018. October 5. What Makes a Good Solar Cell? Adv. Energy Mater. 8(28): 1703385.

[5] Ozturk, S., Aydin, L. and Celik, E. 2018. A comprehensive study on slicing processes optimization of silicon ingot for photovoltaic applications. Sol. Energy. 161: 109–124.

[6] Luque, A. and Hegedus, S. 2011. Handbook of Photovoltaic Science and Engineering. John Wiley & Sons, New York.

[7] Kim, H., Kim, D., Kim, C. and Jeong, H. 2013. Multi-wire sawing of sapphire crystals with reciprocating motion of electroplated diamond wires. CIRP Ann. - Manuf. Technol. 62(1): 335–338.

[8] Kumar, A., Kaminski, S., Melkote, S.N. and Arcona, C. 2016. Effect of wear of diamond wire on surface morphology, roughness and subsurface damage of silicon wafers. Wear 364: 163–168.

[9] Ozturk, S., Kayabasi, E., Celik, E. and Kurt, H. 2018. Determination of lapping parameters for silicon wafer using an artificial neural network. Journal of Materials Science: Materials in Electronics 29(1): 260–270.

[10] Wu, C., Jiang, Z., Fan, W. and Chen, L. 2017. Finite element analysis of multi-wire saw silicon rods with consolidated abrasive diamonds. The Int. J. Adv. Manuf. Technol. 90(1-4): 241–248.

[11] Kayabasi, E., Ozturk, S., Celik, E. and Kurt, H. 2017. Determination of cutting parameters for silicon wafer with a Diamond Wire Saw using an artificial neural network. Sol. Energy 149: 285–293.

CHAPTER 9

Design and Optimization of Permanent Magnet Synchronous Generators

Erhan Nergiz,[1,*] *Levent Aydin,*[2] *Durmuş Uygun,*[1]
Nilay Kucukdogan[3] and *Yucel Celinceviz*[4]

Introduction

Definition and Types of Generators

Generators known as générateur in French are machines which convert mechanical energy into electricity in the literature. Generators are classified in three different types which are direct current generators, asynchronous generators, and synchronous generators. In general, they consist of two main components as stator and rotor.

According to the rotor type, asynchronous generators are divided into two which are doubly-fed asynchronous generators and asynchronous generators with squirrel-cage. Asynchronous machines are electrical machines that convert the alternating current, which is applied to stator windings, into mechanical energy. In asynchronous machines, there is no

---

[1] Department of Research and Development, Aegean Dynamics Corporation, Izmir, Turkey.
Email: durmus.uygun@aegeandynamics.com
[2] Department of Mechanical Engineering, İzmir Katip Çelebi University, Izmir, Turkey.
Email: leventaydinn@gmail.com
[3] Graduate School of Natural and Applied Sciences, İzmir Katip Çelebi University, Izmir, Turkey.
Email: kucukdogan.nilay@gmail.com
[4] Department of Electrical and Electronics Engineering, Kastamonu University, Kastamonu, Turkey.
Email: ycetinceviz@kastamonu.edu.tr
* Corresponding author: erhan.nergiz@aegeandynamics.com

electrical connection between the stator and the rotor, and it operates entirely according to the electromagnetic induction principle. The asynchronous machine needs an AC excitation current. These machines can, therefore, become both self-excited and externally excited machines [1].

Permanent Magnet Synchronous Generators (PMSG) are synchronous machines which are comprised of magnets instead of windings on the rotor. Permanent magnet synchronous generators do not need the mains electricity to start to produce the energy, which is why they are called self-excited generators. Due to their ability to generate energy at low speeds, they are widely used in wind turbine applications [2]. PMSGs are widely used in direct drive wind turbine systems. In direct-drive systems, the generator shaft is directly connected to the shaft of the wind turbine. Therefore, the efficiency is higher and more reliable in direct-drive systems. The advantages of direct-drive permanent magnet generators are a low cost [3, 4], high efficiency, high power density [5, 6], less maintenance, comfortable cooling, and easy control.

Structure of PMSG

Stator and rotor are the main components of the PMSG. The stator is the stationary part of the generator which consists of thin pressed laminated sheets and windings. The other central part is the rotor, where the magnets are located, is the rotating section of the generator [7].

Permanent magnet generators are investigated in three types as in-runner rotor, out-runner rotor, and axial flux according to the rotor structure [8]. In in-runner rotor generators, rotor resides the inside of the generator where the magnets are placed and stator is located on the outer face of the generator and vice versa for the out-runner rotor generators.

The rotor is placed inside, and stator is stated on the outside of the generator for the in-runner PMSGs. The number of poles of in-runner rotors is lower than the out-runner rotors for the same power because the rotor diameter will be decreased. For this reason, when the speed is

Figure 9.1 (1) In Runner PMSG (left) (2) Out-Runner PMSG (middle) (3) Axial Flux PMSG (right).

increased, the power will be decreased for the in-runner rotor generators. Advantages and disadvantages of all generators should be evaluated according to the area of application. In-runner rotor and out-runner rotor generators are also radial flux generators; hence, the rotor turns inside or outside of the stator.

In out-runner rotor PMSGs, the rotor surrounds the stator and magnets are usually located on the inner circumference of the rotor. For this reason, greater rotor diameters and a larger number of poles are allowed in the design. Such generators have advantages and disadvantages compared to their usage areas. In the inner rotor generators, the rotor part where the magnets are located is inside, and the stator part where the windings are located is outside the generator [9].

Axial flux PMSGs are different from the radial flux machines due to the placement of the rotor and stator. In radial flux machines, the rotor rotates inside or outside of the stator, but in axial flux PMSGs both rotor and stator turn axially. This movement can be considered as the rotation of two disks on each other. More than one rotor or stator structure can be designed in axial flux generators, different combinations such as one rotor plus two stators or two rotors plus one stator can be used during the design and production process of axial flux generators. These machines can provide considerable advantages by removing the interconnections of the generator body according to their usage area. Magnets are positioned on the side of the stator for the designed machines in this way. Different versions of these machines are also available. For instance, if a dual-stator and mid-rotating rotor design are considered, the magnets must be placed on both sides of the rotor [10].

Sizing Parameters

There are many variables in PMSG designs that affect efficiency and mass. In this case, it is essential to choose the right variables for generator efficiency. The most important criteria are as follows

- Efficiency
- Load Line Voltage
- Stator Slot Fill Factor
- Total Net Weight
- Cogging Torque
- Winding Voltages
- Winding Currents
- Armature Thermal Load
- Output Power
- Armature Current Density
- Air Gap Flux Density
- Power Factor

The main design parameters are power, frequency, working speed, and load line voltage of generators which affect these criteria. However, these values can be initially calculated analytically according to the desired generation and field of study. In general, the 32 most effective design variables which alter the efficiency of the generator are listed in the following

1. Air Gap	17. Stator Inner Diameter
2. Skew	18. Rotor Inner Diameter
3. Slot Parameter (Bs0)	19. Number of Strands
4. Slot Parameter (Bs1)	20. Magnet Offset
5. Slot Parameter (Bs2)	21. Rotor Thickness
6. Conductors per Slot	22. Slot Parameter (Rs)
7. Stator Steel Type	23. Magnet Type
8. Winding Layers	24. Parallel Branches
9. Stator Outer Diameter	25. Circuit Type
10. Rotor Outer Diameter	26. Wire Size
11. Embrace	27. Magnet Thickness
12. Slot Parameter (Hs0)	28. Number of Poles
13. Slot Parameter (Hs1)	29. Stacking Factor of Stator
14. Slot Parameter (Hs2)	30. Number of Slots
15. Rotor Steel Type	31. Pole Type
16. Type of Winding	32. Stator and Rotor Length

Industrial Applications

From past to present, PM machines have been used in different industrial areas because of their high efficiency, low mass, and volume. Aydin et al. [11] have designed the desired permanent magnet motor to utilize white goods under the required design parameters. High moment-mass ratio, high efficiency, low volume, and mass provide an opportunity in this area [12–14]. Low volume and low mass decrease the cost in the aerial industry which is why Nikbakhsh et al. [15] have used the PM Synchronous Generator in aerospace systems by decreasing the volume of the generator and improve its efficiency. PMSGs' compact sizes, loss reduction, higher power density, optimum efficiency, and reliability features provide advantages, especially in the wind energy sector, which has been heavily invested in, in the last decade [16–18]. Fathabadi [19] has built a prototype car which works with a micro-wind turbine and photovoltaic (PV) module. In this study, there is a permanent magnet synchronous generator state in the wind turbine engine. It was proved that by using wind and solar energy devices together, it adds 19.6 km to the cruising range of the plug-in hybrid electric vehicle. Yousefian et al. [20] worked on a double permanent magnet synchronous generator for wind turbine plants. As it can be seen from the above studies, PMSGs will be considered attractive in many more next-generation studies.

ANSYS Maxwell

ANSYS Maxwell is a leading program for the design and analysis of electric motors and generators, actuators, sensors, converters, and other electromagnetic and electromechanical instruments. When work is done with this program, and the experimental results of these studies are examined, it is observed that the simulation results mostly meet the requirement of real applications. The ANSYS Maxwell program consists of three main parts. These are RMxprt, Maxwell 2D, and Maxwell 3D. In Figure 9.2, a view of ANSYS Maxwell screen is shown.

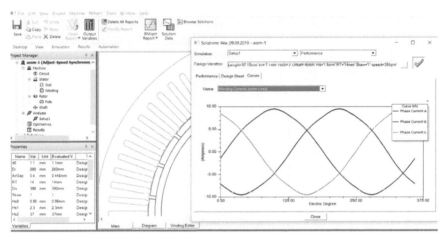

Figure 9.2 ANSYS Maxwell v16.

Rmxprt

RMxprt is a design and optimization software that provides analysis in many different ways, with pre-sizing of electrical machines. Parametric estimation, which is the most significant advantage of the program, can also be done successfully. Thus, calculations involving thousands of conditions can be solved in a short period. Reactions and configurations of the designed machines can be examined in different types of analysis.

Table 9.1 Electric machines which can be solved by Rmxprt.

✓ Adjust Speed ✓ Synchronous Machine	✓ Line-Start Permanent-Magnet Synchronous Motor	✓ Three-Phase Non-Salient Synchronous Machine
✓ Brushless Permanent-Magnet DC Motor	✓ Permanent-Magnet DC Motor	✓ Three-Phase Synchronous Machine
✓ Clow-Pole Alternator	✓ Single-Phase Induction Motor	✓ Universal Motor
✓ DC Machine	✓ Switched Reluctance Motor	✓ Three-Phase Induction Motor
✓ Generic Rotating Machine		

There are 13 types of electric machines that can be solved in this program, and the upcoming table consists of these machines.

PMSG design and optimization can be studied in the "Adjust Speed Synchronous Machine" section. The contents of this section and some of the parameters to be defined are:

Details of Input Variables of Machine

The values set in this part of the program are Number of Poles, Rotor Position (Inner or Outer), Rotor Position (Inner or Outer), Frictional Loss, Windage Loss, Reference Speed, Control Type (DC-PWM-AC) and Circuit Type (Y3-L3-S3-C2-L4-S4).

The number of poles (p) is one of the first parameters that can be set in PMSG design. In synchronous generators, the number of poles depends on the frequency (f), and the generator operating speed (n) that is also reffered to as the reference speed of PMSG. The number of poles can be calculated by using the following formulation:

$$2p = \frac{120f}{n} \tag{9.1}$$

Control Type should be selected as "Alternative (AC)" while designing PMSG for on-grid and off-grid applications. The reference speed value is the desired speed value of the generator to operate at the most efficient point. There exists six different circuit types as:

- Y-Type 3-Phase
- Cross-Type 2-Phase
- Loop-Type 3-Phase
- Loop-Type 4-Phase
- Star-Type 3-Phase
- Star-Type 4-Phase

Stator

The stator parameters are expressed in this section. Besides that, these parameters, which directly affect the efficiency, can be listed as;

- Outer Diameter
- Stacking Factor
- Slot Type
- Inner Diameter
- Steel Type
- Skew Width
- Length
- Number of Slots
- Slot Type

Stator Slot

The most important values affecting the PMSG efficiency are the slot parameters. By selecting these parameters correctly, both the PMSG efficiency can be increased, and the cogging torque can be reduced.

The values that should be known when designing a slot can be listed as follows;

- Hs0
- Hs1
- Hs2
- Bs0
- Bs1
- Bs2
- Rs

All of these parameters belong to the stator slot. During the design process of generator, it should be insisted that Bs0 must be greater than the conductor diameter in terms of ease of winding. Slot parameters are shown in Figure 9.3.

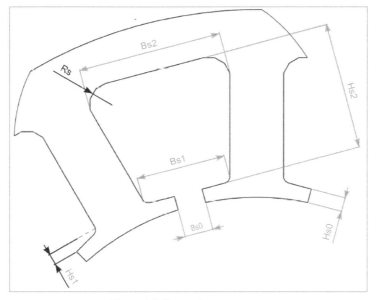

Figure 9.3 Stator slot parameters.

Stator Winding

Winding values are also crucial in PMSG design. For this reason, it is necessary to select these values precisely in order to obtain high efficiency at desired levels such as power and voltage. The parameters to be entered in the winding section are;

- Winding Layers (1 or 2)
- Winding Type (Whole-Coiled or Half-Coiled)
- Parallel Branches
- Wire Size
- Conductors per Slot
- Number of Strands
- Wire Wrap

Rotor

Most of the parameters to be entered in the rotor section are similar to those in the stator section, but the values of these parameters will be different. Rotor inner diameter should be higher than stator outer diameter in out-runner generators. It is vice versa for in-runner generators. Stator and rotor length are the same in all generators except distinctive designs. Rotor parameters can be listed as follows:

- Outer Diameter
- Inner Diameter
- Length
- Steel Type
- Stacking Factor
- Pole Type

Pole

In this section, parameters of the magnet are entered. The shape, dimensions, and material of the magnet directly affects the PMSG efficiency. Embrace, Offset, Magnet Type, and Magnet Thickness are the pole parameters.

Optimetrics

ANSYS Optimetrics is a multifunctional software program which adds parametric, optimization, sensitivity and statistical analysis capabilities to ANSYS HFSS, ANSYS Maxwell and ANSYS Q3D Extractor. Optimetric Analysis consists of different analysis sections, such as Parametric Analysis, Optimization Section, Sensitivity, Statistical Option, and Tuning.

Optimization of PMSG

Recent Studies

Uygun et al. (2016) [21] used the Maxwell RMxprt generator model depending on the ANSYS Simploler external circuit simulator for the PMSG optimization of 0.6 kW power in their work. The co-simulation method was used to analyze the performance of the generator. Analytical calculations were done with ANSYS Rmxprt to find the generator parameters. Experimental test results of the generator which have been designed and produced have been examined. As a result of the research, a PMSG without network connection was designed, optimized, and manufactured. Yıldırız et al. (2009) [22] used a finite element method for the analysis of an axial-flow permanent magnet generator for a small, powerful wind turbine. They used Ansoft Maxwell software for modeling the generator and compared the results of the analysis with those of the prototype machine. In the study, PMSG modeling and electromagnetic analysis were performed for a small wind turbine that could be used in rural areas not reached by electricity. They stated that the closeness of experimental results to FEM results is remarkable. Uygun et al. (2012) [23] have worked on the computer-aided design, analysis, and experimental verification for a PMSG in an external rotor 1 kW power source for small-scale wind turbines. The stator is designed to be skewed to reduce the cogging torque, and the material used is M530-50A silicon steel. The voltages which measured experimentally are higher than the simulated ones by 0.8–1.2%. As a result of the research, it was observed that the intended model is suitable for low-speed wind turbines with low starting torque. Arslan (2014) [24] has worked on a high speed, internal rotor, permanent magnet synchronous generator design. Ansoft Maxwell v14 software was used for modeling of the generator, and two and three-dimensional analyses were done by using the finite element method. After the optimization is completed, the generator is produced, and the test results and the simulation results are compared. The test results examined were found to reflect the simulation output of the Maxwell program. Turkmenoglu et al. (2016) [25] examined the effect of design parameter changes on generator efficiency in synchronous generators with neodymium magnet. In the study, the results were observed by changing the parameters such as the air gap, the type of neodymium magnet, the type of sheet steel used in the stator and the rotor to be compatible with each other, using the design parameters of the neodymium magnet generator, ANSYS (Maxwell, RMxprt). The materials used were changed, and weight and efficiency results were examined. Kahourzade et al. (2013) [26] used GA, finite element, and finite volume methods to perform sizing optimization for axial flow PMSG. It has been attempted to minimize the size of the 1 kW PMSG during operation. It obtained that the simulation results show good agreement with the designed technical specifications. Angle (2016) [27] has worked on the modeling, design, and optimization of permanent magnet synchronous machines. In this study, Maxwell equations were used for solutions, and the techniques of reducing the holding torque were also studied. As a result of the research, it has been observed that the weight can be reduced in the permanent magnets, thus increasing the usability of robot applications requiring lightness. Wang et al. (2016) [28] used an electromagnetic design for hybrid power PMSGs applied in independent power systems, and permanent magnet shape optimization was also performed. The result is high efficiency and a low voltage ripple. FEM and DOE were used in the design process, and simulation outputs were compared with analytical and experimental results. Rastogi et al. (2016) [29] have studied the design, analysis, and optimization of PMSGs. In order to optimize the weight and the losses in the study, $\in$-constrained differential evolution is used together with the gradient-based mutation optimization technique. The Maxwell-Ansys finite element software is used to verify the PMSG design. As a result of the work, the optimal design for 1 kW, 428 rpm PMSG is achieved.

Sindhya et al. (2017) [30] implemented a PMSG design using Multiobjective Interactive optimization. In this study, six objective functions are evaluated in a mixed-integer constrained multiobjective optimization problem. NIMBUS and NSGA-II optimization methods were used to obtain the solutions. After the process, the decision-maker (DM) provided the optimal result for the design. They stated that interactive multi-objective optimization methods could be used for different electrical machine topologies and different models. Ishikawa et al. (2017) [31] optimized the rotor structure in PMSGs for higher output power and less magnet volume utilization by using the GA cluster and cleaning method. The PMSG obtained by this topology optimization was used in the hybrid vehicle Prius. As a result of this work, the output power is increased by 11.6%, and the magnet volume is reduced by 21.8%. Bobon et al. (2017) [32] studied the calculation of nonlinear model parameters based on power rejection tests using PMSG using gradient optimization algorithm. The coefficients of the function in the mathematical models are determined by the least-squares method. In order to minimize the objective function in the study, the Newton gradient optimization algorithm is used. As a result of the work, the final parameter results for 200 MW and 50 MW turbogenerators (synchronous generators with a cylindrical rotor) are estimated, and waveform comparisons are measured and calculated.

While the values of the parameters such as generator efficiency, power, and line voltage are calculated, the methods used in the design process can vary. There exist too many parameters that affect the final design of PMSG. Many of these parameters are described in section RMxprt. If an efficient PMSG design is desired, the parameters must be selected correctly. It is not correct to focus only on the efficiency result during design. Other constraints (cogging torque, stator slot fill factor, weight, etc.), need to be also considered. As it is seen in this section, many methods have to be used to PMSG design and optimization.

Definition of Problem

In this study, the modeling and optimization of 4 kW power generator used in medium-sized wind turbines were performed. Experimental data was obtained from Cetinceviz's doctoral thesis (2017) [33], and the required data set was obtained from Maxwell Rmxprt program.

Objectives and Constraints

The main purpose of this study is to maximize the efficiency of the PMSG. While maximizing the efficiency of the designed generator, the operating conditions are given as follows:

Cogging Torque < 0.01 (Nm), 3800 < **Output Power** < 5000 (Watt), 350 < **Load Line Voltage** < 420 (Volt)
55 < **Stator Slot Fill Factor** < 70(%), **Total Net Weight** < 55 (kg)

These conditions are considered as the constraint functions of the optimization problem. Thus, the problem consists of six functions, one objective and, five constraints, and then particular models are created for each function.

Design Variables and Constant Parameters

In this study, the aim is to increase the efficiency of the generator designed in the reference study by keeping some of the parameters constant. The data required for optimization was obtained using Parametric Analysis in the Adjust Speed Synchronous Machine section of the ANSYS Maxwell RMxprt program, and machine type is a generator. Reference generator and constraint parameters are shown in Table 9.2.

In this study, 19 design variables (see Table 9.3) that directly affect the objective and constraint functions were determined. When determining these variables, their effect on efficiency and other parameters was taken

Table 9.2 Constraint parameters.

Number of Poles = 24	Number of Slots = 72	Frequency = 50 Hz	Reference Speed = 250 rpm
Rotor Position: Inrunner	Control Type: Alternating Current	Circuit Type: Y3	Stacking Factor = 0.95
Stator Material: M19_24G	Rotor Material: st-37	Magnet Material: NdFe35	Slot Type: 3rd
Pole Type: Surface Located	Load Type: Independent Generator	Winding Type: Whole Coiled	Winding Step: 3
Power Factor: 1	Winding Layers: 1	Parallel Branches: 1	

Table 9.3 Design variables and their symbols used in the models.

• Air Gap (AG)	• Slot Parameter 1 (Bs0)	• Slot Parameter 2 (Bs1)	• Slot Parameter 3 (Bs2)
• Conductors per Slot (CS)	• Stator Inner Diameter (Di)	• Stator Outer Diameter (Do)	• Embrace (E)
• Slot Parameter 4 (Hs0)	• Slot Parameter 5 (Hs1)	• Slot Parameter 6 (Hs2)	• Stator and Rotor Length (L)
• Number of Strands (NS)	• Offset (O)	• Rotor Thickness (RT)	• Slot Parameter 7 (Rs)
• Skew (S)	• Wire Size (d0)	• Magnet Thickness (lm)	

into consideration. After determining the design variables, their effects on other functions at their maximum and minimum points were examined.

Acquisition of Data

Firstly, while each variable is in average value, it is solved in the Maxwell program, and the results of functions are examined. Then, solutions were made for the maximum and minimum values for each variable respectively and, its effect on functions was examined. Thus, it was attempted to determine which variable is more effective for each function. After that, it was calculated which variable affects each function more. The flow of this process is given below:

- The percentage change (PC) of the variable from minimum to maximum was calculated;

$$PC = \left(\frac{x_{max} - x_{min}}{x_{max}} \right) * 100 \tag{9.1}$$

- This value was indicated in the table as a percentage of change.
- Then, the minimum value is subtracted from the maximum one obtained as a result of the function, and, the result was added to the table as the change value.
- Finally, the function changes according to the percentage change rate are calculated with the following formula. These were added to the table in order of percentage change:

$$\frac{change * 100}{PC} \tag{9.2}$$

After these steps, the effect of variables on the rate of change was examined, and, the variables with more effect were given higher value.

In order to obtain "Efficiency Model" data, values were given to variables in Maxwell parametric analysis and an efficiency result was achieved for 30720 different combinations. Maxwell parametric analysis took about 15 days to generate all the data (HP Z1 8 Core Work Station, RAM 32 GB).

Mathematical Models

In this section, we aim to develop separate models for efficiency, cogging torque, output power, load line voltage, and stator slot fill factor and weight parameters as a function of generator design variables. As highlighted in the previous sections, the efficiency and other parameters depend on the generator design variables.

Mathematical models which were prepared in multiple non-linear regression types by using data obtained from the Maxwell program have been developed to define optimum values of PMSG parameters. The gathered information after this process was used to optimize the efficiency and determine the effect of the parameters on the objective functions.

In the preparation of the models, standard multiple non-linear regression analysis and Artificial Neural Network (ANN) methods were used together in hybrid form. Therefore, the data is randomly divided into two parts of 80% and 20% for each model. Regression models were prepared using 80% data and, R^2 values representing the accuracy of the model were calculated using 20% data. The models created with 80% data were named "training," and the validation results with 20% data were called "testing." The Wolfram Mathematica program was used to construct models, compare results and, calculate R^2s. The general form of the regression equation used for all models is as follows:

$$\mathbf{y} = [a0 + (\textstyle\sum_1^{19} a_i\, x_i)^2]/\, [b0 + (\textstyle\sum_1^{19} b\, x_i)^2] \tag{9.3}$$

In this function, x and y represent the encoded values of the design variables and the output value for the models, respectively. The values of a_0, a_1, a_2, ..., b_0, b_1, b_2 express the coefficients of the model, and they were determined using the least-squares technique. R^2 and R^2_{adj} values were calculated to determine the accuracy of the model for training and testing results, respectively and given in Table 9.4.

Definition of Optimization Problem

In this study, one nonlinear discrete optimization problem was solved. The limitations of the solved problem were shown in Table 9.5.

Results and Discussion

The values of the design variables are shown in Table 9.6, which were obtained from the created optimization problem solutions in Mathematica. Besides that, it was found that the efficiency reached 99% according to the results obtained. The values of the design variables were entered into the Maxwell program and, efficiency value was found to be 95.72%. The 4 kW efficiency value is quite high for a PMSG. In addition to these variables and efficiency value, initially targeted constraints were examined. The simulation, measurement, and analytical results for the optimized PMSG for the reference study [33] were shown in Table 9.7. In this study, it was aimed to improve the efficiency for the same PMSG. The test results were very close to the analytical and simulation results. The created models within the scope of this study were solved in the optimization problems shown in Table 9.5.

Table 9.4 The R^2 and R^2_{adj} results for the efficiency model.

R^2	Efficiency
$R^2_{training}$ (80%)	0.950
$R^2_{testing}$ (20%)	0.949
R^2_{adj}	0.950

Table 9.5 Definition of optimization problem including constraints.

Objective: Max Efficiency
Constraints and Bounds: 0.1<AG<2, 2<Bs0<7, 3<Bs1<12, 3<Bs2<15, 30<CS<50, 260<Di<360,   330<Dout<430, 0.6<E<1.1, 0.1<Hs0<3, 0.1<Hs1<5, 25<Hs2<50, 50<L<120,   2<NS<6, 1<O<110, 10<RT<30, 0<Rs<4, 0.1<S<2, 0<d0<3, 5<lm<10,   60<(((3.14*(d0/2)^2*NS*CS*120))/((((Bs1+Bs2))/2)*Hs2)))<70,   Bs0<Bs1,((2*Hs2)+Di)<Dout
Design Variables: AG, Bs0, Bs1, Bs2, CS, Di, Dout, E, Hs0, Hs1, Hs2, L, NS, O, RT, Rs, S, d0, lm

Table 9.6 Optimum design results based on the proposed modeling-optimization process.

Efficiency% = 99
AG = 0.448, Bs0 = 4.3, Bs1 = 6.5, Bs2 = 6.2, CS = 38., Di = 280, Dout = 390,   E = 0.91, Hs0 = 0.95, Hs1 = 2.3, Hs2 = 37, L = 85.16, NS = 3, O = 40, RT = 23,   Rs = 1.5, S = 1, d0 = 1.1, lm = 7.1

Table 9.7 Comparison between the results of the reference study and those of the present study.

Parameters	Analytical [33]	Simulation [33]	Experimental [33]	Present
Efficiency (%)	92.9	93.07	93	95.72
Cogging Torque (Nm)	–	0	–	0
Output Power (Watt)	4791	4885.87	4740	4561.94
Load Line Voltage (Volt)	415.3	421	442.89	408.9
Stator Slot Fill Factor (%)	–	68.46	–	56.16
Total Net Weight (kg)	–	35	–	46.8

In the evaluation of the results, firstly the design variable results obtained by using Wolfram Mathematica were entered into ANSYS Ansoft Maxwell program, and Maxwell results (see Table 9.6) values were obtained. Efficiency is the objective function in the optimization problem, and it is aimed to maximize it.

The results of the final models were examined and illustrated in the Maxwell program. When the graphs are considered, it is seen that the model is well prepared in terms of electrical and magnetic aspects. Firstly, the winding voltages and currents of the generator depending on the Electric Degree are examined. The behaviors of the winding voltages and the currents versus the Electric Degree are shown in Figures 9.4 and 9.5, respectively.

The primary purpose of the study is to increase the efficiency of the generator. Therefore, the results of the speed-efficiency graph of the final generator shown in Figure 9.6 in the range of 50–300 rpm are examined. When the graph is examined, it is seen that the efficiency is a decreasing curve depending on the speed. The figure shows that the generator's efficiency at the reference speed 250 rpm is 95.72%. When the speed approaches 300 rpm, the efficiency reaches 97%. Also, as the output power is increased from 1 kW to 4 kW, the efficiency curve also moves upwards.

However, as the winding voltage and current go up with the increase in speed poses a risk. For this reason, the current density should be examined when the generator is operating at high speeds. For example, if the current density is greater than 6, the copper losses in the windings will increase, and the generator will cause high-intensity heating. The current density curve of the optimized generator at different powers depending on the speed is shown in Figure 9.7. When the figure is examined, it is seen that the generator can operate smoothly even at high speeds in terms of current density. Even at high power and speed, the current density is less than 3 A/mm². It has been observed that the current density increases when the output power increases with speed.

The line voltage results for the different speed and power values are shown in Figure 9.8. When the graph is examined, the voltage remains at 350–420 V at 250 rpm, which is the reference speed. The output voltage increases linearly depending on the speed and gives close values at different powers. The obtained generator model can provide the desired voltage at the reference speed level. Furthermore, the current density directly acting on the generator heating remains within the desired range.

Figure 9.9 shows the general diagram of the optimized generator. In the graph, the generator's torque, efficiency, and line voltage parameters are examined depending on the output power at 250 rpm. The generator torque shows a linear increase depending on the output power. When the efficiency values at different powers are examined, the generator draws a smooth curve that rises continuously from 1 kW up to 5 kW. In other words, the generator starts from 1 kW at the reference speed and starts working in the productive zone. This wide range of high efficiency is perfect for PMSGs. In this study, since the efficiency optimization of a 4 kW output PMSG is aimed, the efficiency graph is such that the results

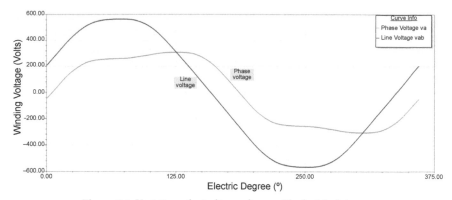

Figure 9.4 Variation of winding voltage with electrical degree.

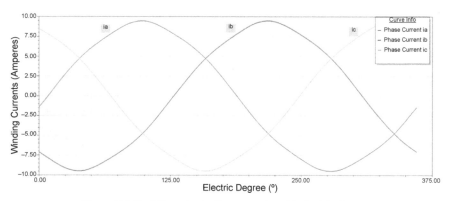

Figure 9.5 Variation of winding currents with electric degree.

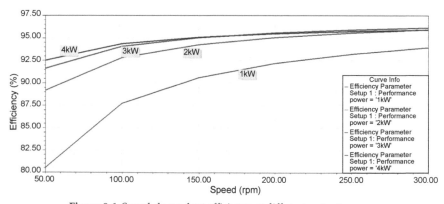

Figure 9.6 Speed-dependent efficiency at different output powers.

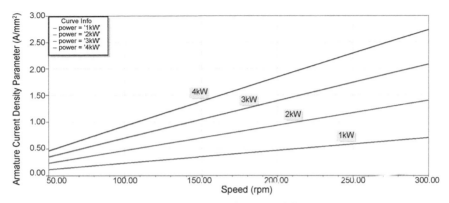

Figure 9.7 Speed-dependent current density at different output powers.

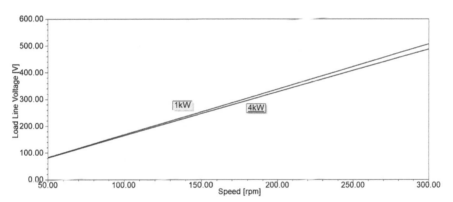

Figure 9.8 Speed-dependent load line voltage at different powers.

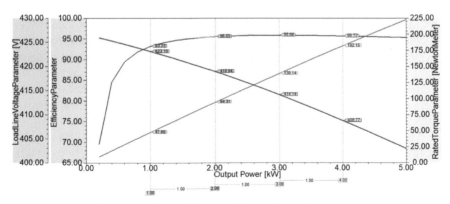

Figure 9.9 General diagram of the optimized generator.

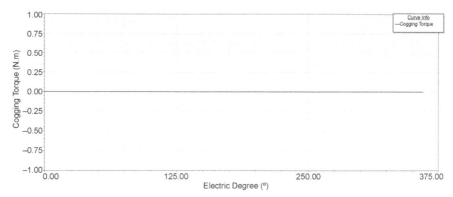

Figure 9.10 Cogging torque related to electric degree.

are in the desired range. Besides, line voltage decreases due to the increase in generator power. It is seen that the output voltage of the generator is 408.77 V at 250 rpm of rate and 4 kW of output power.

In this study, it was also aimed to minimize the cogging torque while increasing efficiency. Figure 9.10 shows the electrical torque-dependent cogging torque graph of the optimized generator model. When the graph is examined, it is seen that the cogging torque is zero at every angle. For a generator of this scale, cogging torque of zero or close to zero means that it can produce energy even at very low wind speeds. The cogging torque is a critical parameter for power generation at low wind speeds.

Future Aspects

As is known, the advancement in permanent magnet technology provides higher energy outputs from different renewable energy sources such as wind and river power plants. For this reason, the works in this field were accelerated in the 20th century, and many new designed generators were designed with this infrastructure. If the technology in permanent magnets continues to develop by providing higher energy and magnetic flux densities produced by different families of permanent magnet materials. Permanent magnet materials and PMSGs will in the future produce effective energy solutions from various natural energy sources.

In this study, the 4 kW PMSG is discussed, and its efficiency is optimized. One of the future studies can be the production of this optimized generator and the examination of the test results. Test and optimization results can be compared, and the accuracy of the methods can be examined. If the test results are successful, it is aimed to determine a new generator model according to the needs of the market and to make more optimizations by considering more design variables.

References

[1] Kumawat, R.K., Chourasiya, S., Agrawal, S. and Paliwalia, D.K. 2015. Self excited induction generator: A review. International Advanced Research Journal in Science, Engineering and Technology (IARJSET), National Conference on Renewable Energy and Environment (NCREE-2015) Vol. 2, Special Issue 1.

[2] Anvari, B., Li, Y. and Toliyat, H.A. 2015. Comparison of outer rotor permanent magnet and magnet-less generators for direct-drive wind turbine applications, 181–186. The Proceedings of the Electric Machines & Drives Conference (IEMDC), Coeur d'Alene, ID, USA. IEEE.

[3] Jang, S.M., Seo, H.J., Park, Y.S., Park, H. and Choi, J.Y. 2012. Design and electromagnetic field characteristic analysis of 1.5 kW small scale wind power generator for substitution of Nd-Fe-B to ferrite permanent magnet. IEEE T. Magn. 48(11): 2933–2936.

[4] Potgieter, J.H.J. and Kamper, M.J. 2012. Torque and voltage quality in design optimization of low-cost non-overlap single layer winding permanent magnet wind generator. IEEE T. Ind. Electron. 59(5): 2147–2156.

[5] Wang, H., Wang, Y., Liu, X., Coa, G. and Fan, J. 2011. Design and performance evaluation of compound permanent magnet generator with controllable air-gap flux. IEE Proc-B 5(9): 684–690.

[6] Tapia, J.A., Pyrhönen, J., Puranen, J., Lindh, P. and Nyman, S. 2013. Optimal design of large permanent magnet synchronous generators. IEEE T. Magn. 4(1): 642–650.

[7] Elosegui, I., Martinez-Iturralde, M., Rico, A.G., Florez, J., Echeverría, J.M. and Fontan, L. 2007. Analytical design of synchronous permanent magnet motor/generators, 1165–1170. The Proceedings of the International Symposium on Industrial Electronics, Vigo, Spain. IEEE.

[8] Madani, N. 2011. Design of a Permanent Magnet Synchronous Generator for a Vertical Axis Wind Turbine. M.S. Thesis, KTH Royal Institute of Technology, Stockholm, Sweden.

[9] Li, Y., Zhao, J., Chen, Z. and Liu, X. 2014. Investigation of a five-phase dual-rotor permanent magnet synchronous motor used for electric vehicles. Energies 7(6): 3955–3984.

[10] Kinnunen, J. 2007. Direct-On-Line Axial Flux Permanent Magnet Synchronous Generator Static and Dynamic Performance. Ph.D. Thesis, Lappeenranta University of Technology, Lappeenranta, Finland.

[11] Aydın, M., Ocak, O. and Albas, G.C. 2012. Permanent magnet synchronous motors used in direct-drive home appliance applications. EMO Bilimsel Dergi. 2(3): 7–11.

[12] Sebastain, T. and Slemon, G.R. 1987. Operation limits of an inverter-driven permanent magnet motor drives. IEEE T. Ind. Appl. 23(2): 327–333.

[13] Sebastian, T., Slemon, G.R. and Rahman, M.A. 1986. Design considerations for variable speed permanent magnet motors, 1099–1102. Proceedings of International Conference on Electrical Machines (ICEM), Munich, Germany.

[14] Boules, N. 1984. Prediction of no-load flux density distribution of permanent magnet machines. IEEE IEEE T. Ind. Appl. 21(3): 633–643.

[15] Nikbakhsh, A., Gholamian, S.A. and Hoseini, S.M. 2014. Optimal design of PM synchronous generator for aerial industries power supply using cuckoo algorithm. Int. J. Res. Ind. Eng. 3(3): 49–68.

[16] Nagorny, A.S., Dravid, N.V., Jansen, R.H. and Kenny, B.H. 2005. Design Aspects of a High-Speed Permanent Magnet Synchronous Motor/Generator for Flywheel Applications, 635–641. International Electric Machines and Drives Conference. San Antonio, USA. IEEE.

[17] Jones, R.I. 1999. The more electric aircraft: The past and the future, 1–4. The Proceedings of the Electrical Machines and Systems for the More Electric Aircraft, London, UK. IET.

[18] Gieras, J.F. 2013. Electric Power System of Tu–154M Passenger Aircraft. Prz. Elektrotechniczn. 3: 300–307.

[19] Fathabadi, H. 2018. Utilizing solar and wind energy in plug-in hybrid electric vehicles. Energ. Convers. Manage. 156: 317–328.

[20] Yousefian, H. and Kelk, H.M. 2018. A unique optimized double-stator permanent-magnet synchronous generator in high-power wind plants. Energy. 143: 973–979.

[21] Uygun, D., Cetinceviz, Y. and Bal, G. 2016. Optimization study on a 0.6 kW PMSG for VAWTs and determination of open and short circuit performances by using external circuit method. Int. J. Hydrogen Energ. 41: 12535–12545.

[22] Yildiriz, E. and Aydemir, M.T. 2009. Küçük güçlü bir rüzgar jeneratöründe kullanim için eksenel akili bir sürekli miknatisli motorun analizi, tasarimi ve gerçekleştirilmesi. J. Fac. Eng. Archit. Gaz. 24(3): 535–531.

[23] Uygun, D., Ocak, C., Cetinceviz, Y., Demir, E. and Gungor, Y. 2012. CAD-based design, analysis and experimental verification of an out-runner permanent magnet synchronous generator for small scale wind turbines, 179–183. 11th International Conference on Environment and Electrical Engineering, Venice, Italy, IEEE.

[24] Arslan, S. 2014. Yuksek hızlı kalıcı mıknatıslı generator tasarımı ve analizi. M.S. Thesis, Gazi University, Ankara, Turkey.

[25] Turkmenoglu, V., Arslan, S., Yusufoglu, A., Fenercioglu, A. and Atasorkun, M. 2016. Neodyum mıknatıslı senkron generatörlerde tasarım parameter değişikliklerinin generator verimine etkisi. Adv. Manuf. Sci. Technol. 5(2): 101–111.

[26] Kahourzade, S., Mahmoudi, A., Gandomkar, A., Rahim, N.A., Ping, H.W. and Uddin, M.N. 2013. Design optimization and analysis of AFPM synchronous machine incorporating power density, thermal analysis, and back-emf THD. Prog. Electromagn. Res. 136: 327–367.

[27] Angle, M.G. 2016. Modeling, Design, and Optimization of Permanent Magnet Synchronous Machines. Ph.D. Thesis, Massachusetts Institute of Technology, Cambridge, USA.

[28] Wang, H., Qu, Z., Tang, S., Pang, M. and Zhang, M. 2016. Analysis and optimization of hybrid excitation permanent magnet synchronous generator for stand-alone power system. J. Magn. and Magnetic Mater. 436: 117–125.

[29] Rastogi, S., Kumar, R.R. and Singh, S. 2016. Design, analysis and optimization of permanent magnet synchronous generator, 1–5. International Conference on Power Electronics, Drives and Energy Systems, Trivandrum, India, PEDES.

[30] Sindhya, K., Manninen, A., Miettinen, K. and Pippuri, J. 2017. Design of permanent magnet synchronous generator using Interactive Multiobjective Optimization. Transactions on Industrial Electronics. 64(12): 9776–9783.

[31] Ishikawa, T., Watanabe, T. and Kurita, N. 2017. Effect of cleaning level on topology optimization of permanent magnet synchronous generator. J. Ind. App. 6(6): 416–421.

[32] Bobon, A., Nocon, A., Paszek, S. and Pruski, P. 2017. Determination of synchronous generator nonlinear model parameters based on power rejection tests using a gradient optimization algorithm. B. Pol. Acad. Sci-Tech. 65(4): 479–488.

[33] Cetinceviz, Y. 2017. Küçük Ölçekli Rüzgar ve Nehir Santralleri İçin 4 kW Gücünde Kalıcı Mıknatıslı Senkron Generatör Tasarımı ve Gerçekleştirilmesi. PhD. Thesis, Karabuk University, Karabuk, Turkey.

CHAPTER 10

Design Optimization of an Offshore Wind Turbine

H Irem Erten,[1] Levent Aydin,[2] Selda Oterkus,[3,]* Erkan Oterkus[3] and H Seçil Artem[1]

Introduction

Wind energy technology shows an increase in the recent reports of wind power capacity, expectations of market expansion and international research projects. Wind turbines are one of the essential devices used in the wind energy sector [1]. In recent years, wind turbines have started to be installed at offshore locations. Therefore, there is a rapid progress of offshore wind turbine related research in the literature. For example, the concept of offshore wind turbines was first proposed by Honnef in the 1930s [2]. In Europe, the first studies associated with offshore wind energy were published in the late 1970s. One of the first studies associated with offshore wind energy was on the feasibility and construction of offshore wind turbines [3]. The first single offshore wind turbine was installed in Nogersund, Sweden, in 1990 [4].

According to their platforms, offshore wind turbines can be divided into three primary concepts: (i) ballast stabilized spar-buoy, (ii) mooring line stabilized tension leg platform (TLP), and (iii) buoyancy stabilized barge platform. Each type of offshore wind turbine has its upsides and

[1] Izmir Institute of Technology, Department of Mechanical Engineering, İzmir, Turkey.
  Emails: erteniremm@gmail.com; secilartem@iyte.edu.tr
[2] İzmir Katip Çelebi University, Department of Mechanical Engineering, İzmir, Turkey.
  Email: leventaydinn@gmail.com
[3] University of Strathclyde, PeriDynamics Research Centre, Department of Naval Architecture, Ocean & Marine Engineering, Glasgow, Scotland.
  Email: erkan.oterkus@strath.ac.uk
* Corresponding author: selda.oterkus@strath.ac.uk

its own limitations. Ballast stabilized spar-buoy platforms are deepwater structures. They are used for gas and oil production in water depths of about 100 meter [5]. TLPs have been exclusively used for platforms for drilling and production. This platform is a vertically moored compliant. The platform, which has excess buoyancy, is moored by taut mooring lines called tendons (tethers). All structures of the platform are vertically restrained to prevent rotational (pitch and roll) and vertical (heave) motions. Besides that, they are compatible in the horizontal direction allowing for lateral motions [5, 6]. The buoyancy stabilized floating offshore wind turbine (FOWT) platform includes a barge base with catenary mooring lines. The lines create a curved shape to increase the durability of the anchors. The main advantage of such a barge-based mooring principal is its comparatively low anchor cost [7].

Recently, structural and dynamic performances of FOWTs in the turbulent wind flow condition was investigated by Li et al. [8]. The individual effects of turbulence density, wind shear, and their effect on the structural and dynamic responses of FOWTs were investigated using the aero-hydro-servo-elastic design code "FAST" (Fatigue, Aerodynamics, Structures and Turbulence) studied by the National Renewable Energy Laboratory (NREL) [9]. The dynamics of a TLP-FOWT by using experimental and numerical methods were investigated by Oguz et al. [10]. In this chapter, the benefits of TLP structures were demonstrated by using time-varying aerodynamic forces obtained by the wind turbine during a physical test and by using a fully coupled FAST code developed by NREL. The results of this study demonstrated the benefits of such TLP structures which are crucial to obtaining a super high power output. Martin et al. [11] described observational research to compare three FOWT types; Tension line platform, semi-submersible and spar buoy. The front edge of the model blade was preternaturally roughed to increase the aerodynamic performance of the rotor. It has been noted that this method might result in unbalanced wind turbine behavior. Therefore, these methods are not only used for fine-tuning but rather for a complete solution for improving the aerodynamic performance of offshore wind turbines. It is proposed that the best technical approach is to redesign the rotor and use other methods to adjust finely the model thrust forces. Besides this, several studies include the optimization of FOWT foundations.

Bachynski and Moan studied [12] the five different parametric single-columns of TLP wind turbines which have been designed and researched with four different wind wave circumstances by using the Riflex, Simo, and Aerodyn numerical tools to determine the platform movements and structural loads on the components and tendons of the turbine. Rodriguez et al. [13] have studied the hydrodynamics performance of a TLP to support a wind energy turbine. They present the experimental setup and results from decayed experiments, irregular wave responses, regular wave movement, tendon loads and accelerations. In the study, experimental outcomes with present in-house numerical simulations and other outcomes which are obtained from literature were compared. Their experimental outcomes indicated that the damping values and natural periods are similar to given values in the literature.

It is essential to consider the systematic design methods and performance of marine structures. Therefore, engineering optimization techniques have been developed for many engineering problems which are related to marine systems. Engineering optimization is a subject which uses optimization techniques to obtain the best design and performance evaluation in engineering systems quickly. Optimal design problems in marine systems include hydrodynamics and structural problems (superstructures, substructures, and offshore structures) to increase their performance. Hence, there are several methods to specify the performance and configuration of the FWOT [14]. For instance, Ramachandran and Vasanta [15] introduced a numerical model for TLP to derive the platform and hydrodynamic forces and applied an advanced aero-elastic code called Flex5, to calculate the hydro-aero-servo-elastic loads and responses on the wind turbine and floater. In addition to these, Aktaş et al. [16] developed hydrodynamic models for two different types of platforms which are the buoyancy stabilized barge platform and ballast stabilized spar-buoy using ANSYS AQWA and their results were compared with other simulation models and physical experiments in the literature. According to their study, the results show that ANSYS-AQWA model can be used in the hydrodynamic modeling of FOWTs. Yang et al. [17] analyzed reliability-based design optimization of the tripod sub-structure of the offshore wind turbines with dynamic limitations by using the kriging method, and the structure weight was accepted as the objective function. The reliability of the structure was specified via Monte Carlo simulations. The present results indicate that the recommended method may submit a dependable design with less weight and better dynamic performances. Damiani [18] presented studies related to the design of offshore wind turbine (OWT) towers. This study includes loading scenario identification, topology selection, computer-aided engineering, structural reliability, standards and codes, structural dynamics, integrity limitations, equations of simplified design, and integrated systems engineering. According to their investigations, floating platforms can also assist in reducing loads in the tower if the control system is sufficiently designed. These viewpoints combined with active damping units, intelligent controls, and new materials contributed to the research in the reduced cost of energy. In maritime conditions, unstable vibrations can occur for the barge-type OWT. Yang et al. [19] proposed an amendable method involving a Tuned

Mass Damper (TMD) in OWT platforms to show the advancement of the dynamic structural performance by using Lagrange's equations in order to estimate a restricted degree of freedom (DOF) mathematical model.

Another significant performance of engineering systems is related to their geometries. Efficient operation of an OWT structure requires a favorable degree of shape optimization. Therefore, the geometry of the engineering systems should be optimized for the best alternative [14].

Optimization techniques can be categorized as traditional and non-traditional (Approximation Methods). Traditional techniques are analytical, and it is only proper to obtain the optimum solutions of continuous and differentiable functions. On the other hand, non-traditional techniques are proper for models which have the continuous-discrete and discrete domain. Some of the non-traditional optimization tools are Differential Evolution (DE), Genetic Algorithms (GA), Simulated Annealing (SA) and others. These types of non-traditional optimization techniques are stochastic. Design and optimization problems of FOWT contain sophisticated and high non-linear functions. In these circumstances, the stochastic optimization methods such as Genetic Algorithms (GA), Differential Evolution (DE), and Simulated Annealing (SA) are proper for these type of problems [20–22].

In contrast, there is very little research about the optimization process, including stochastic optimization methods in FOWT problems. For instance, Gentils et al. [23] developed a structural optimization model for offshore wind turbine support structures through Finite Element Analysis (FEA) and Genetic Algorithm, minimizing the support structure mass with multi-criteria constraints in contrast to available optimization models for offshore wind turbine support structures, the presented model is an integrated structural optimization model which optimizes the support structure components at the same time. It indicated that the presented structural optimization model is capable of effectively determining the optimal design of offshore wind turbine support structures, which considerably play a role on their design efficiency.

In this chapter, the substructure of the NREL 5 MW TLP type of OWT is taken into consideration. The primary aim is to determine optimal TLP-type of FOWT's substructure for maximum performance considering the hydrodynamics productivity. Hence, best mathematical models for optimal design are constructed through a commercial software Mathematica. The accuracy of the constructed models is checked through R^2 training and testing values. After testing the reliability of the models, it is aimed to minimize the outputs (nacelle acceleration and line tension). To optimize the parameters, Mathematica was used for the best design. This software also involves stochastic methods which are Random Search (RS), Simulated Annealing (SA), and Differential Evolution (DE), which are described in Chapter 2, and a deterministic one Nelder-Mead (NM).

Problem Definition

The substructure of NREL 5 MW baseline TLP type of OWT is taken into consideration. Figure 10.1 represents the design model and design parameters. In this figure, R_C represents the column radius, R_B represents the radius of the cylinder, H represents the cylinder height, and S represents the submerged depth [14].

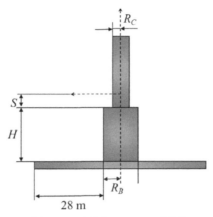

Figure 10.1 Substructure of TLP.

Wind turbine specifications are demonstrated in Table 10.1. The submerged depth (S), the height of the cylinder (H), column radius (R_C), cylinder radius (R_B), nacelle acceleration (NA) and line tension (LT) data is taken from the study conducted by Lee et al. [13]. In their study, the west coast of Korea near Chilbal-Island was taken into consideration for the ambient conditions present with a wind speed of 11.4 m/s, wave height of 4.11 meter and a wave period of 12.49 sec and the JONSWAP spectrum was used [14]. Tables 10.2 and 10.3 represent the line tension and nacelle acceleration generated by using commercial code (AQWA). In Table 10.2, case 9 represents the base design values, which is the NREL 5 MW TLP-type wind turbine [14].

Regression analysis is a proper choice tool to estimate the relationships among the parameters which affect the engineering process. In the literature, there are many types of regression models such as Logistic, Linear, Nonlinear and Stepwise to define the many phenomena. When the number of inputs is more than one, the multiple-regression method is appropriate for the present data. Furthermore, if the physical phenomenon is not linear in the sense of parameters, nonlinear multiple regression models can be used to foresee the phenomenological response of the process. Therefore, in this study, multiple nonlinear regression analysis is used [24].

Table 10.1 Specification of wind turbine [14].

Constant	Value
Wind speed	11.4 m/s
Turbine moment	72,000 Nm
Turbine trust	800 kN
Wind turbine	NREL 5 MW Baseline Wind turbine

Table 10.2 Training data [14].

Case	S	R_B	H	Nacelle Acceleration	Line Tension
1	0.500	0.500	0.500	0.780	0.500
2	1.000	0.500	0.500	0.590	0.559
4	1.000	0.500	0.750	0.551	0.555
5	0.500	0.500	1.000	0.670	0.586
6	1.000	0.500	1.000	0.500	0.576
7	0.500	0.750	0.500	0.913	0.640
9	0.500	0.750	0.750	0.832	0.660
10	1.000	0.750	0.750	0.607	0.695
11	0.500	0.750	1.000	0.754	0.714
12	1.000	0.750	1.000	0.553	0.773
14	1.000	1.000	0.500	0.714	0.798
15	0.500	1.000	0.750	0.899	0.688
16	1.000	1.000	0.750	0.649	0.706
17	0.500	1.000	1.000	0.790	1.000
18	1.000	1.000	1.000	0.589	0.940

Table 10.3 Testing data [14].

Case	S	R_B	H	Nacelle Acceleration	Line Tension
3	0.500	0.500	0.750	0.723	0.587
8	1.000	0.750	0.500	0.665	0.667
13	0.500	1.000	0.500	1.000	0.770

Optimum design can be obtained by the following steps:

1) Neuro-regression modeling
2) Boundedness of the model
3) Optimization

1) **Neuro-regression modeling:** Nacelle acceleration and line tension are two different engineering system parameters. Tables 10.2 and 10.3 represent rescaled training and test data, respectively [14]. Training data consists of 80% of the data. The testing data is used to control the accuracy of the proposed model. After this separation, four different numerical models (see Table 10.4) have been proposed for two engineering parameters. As presented in Tables 10.4–10.7, R^2 training, R^2 training-adjusted R^2 testing values and coefficients have been calculated for each model by the help of Mathematica software.

In the models, X_1, X_2 and X_3 refers to the submerged depth (S), the cylinder radius (R_B) and the height of the cylinder (H) respectively.

2) **Boundedness of the model:** As presented in Tables 10.8 and 10.9, the maximum and minimum values of the proposed functions within the limitation of the design variables are calculated whether the models are bounded or not. After evaluating these values, it is concluded that the models are appropriate for this engineering problem.

For the line tension, when the exponential value of the inputs in the same model structure is changed, and as they give distant testing values to each other, four different models have been proposed. Then, maximum and minimum values for line tension have been calculated by using Mathematica.

Table 10.4 Models for line tension.

Models No.	Models	R^2 Training	R^2 Adjusted	R^2 Testing
1	$\sum_{k=1}^{3}\sum_{n=1}^{3}\sum_{m=1}^{5} \dfrac{a_{m+n+k} (1 + Sin(X_1^m) + Sin(X_2^n) + Sin(X_3^k))}{b_{m+n+k} (1 + Sin(X_1^m) + Sin(X_2^n) + Sin(X_3^k))}$	0.994	1.00	0.843
2	$\sum_{k=1}^{4}\sum_{n=1}^{4}\sum_{m=1}^{5} \dfrac{a_{m+n+k} (1 + Sin(X_1^m) + Sin(X_2^n) + Sin(X_3^k))}{b_{m+n+k} (1 + Sin(X_1^m) + Sin(X_2^n) + Sin(X_3^k))}$	0.994	1.00	0.878
3	$\sum_{k=1}^{3}\sum_{n=1}^{3}\sum_{m=1}^{4} \dfrac{a_{m+n+k} (1 + Sin(X_1^m) + Sin(X_2^n) + Sin(X_3^k))}{b_{m+n+k} (1 + Sin(X_1^m) + Sin(X_2^n) + Sin(X_3^k))}$	0.994	1.01	0.913
4	$\sum_{k=1}^{2}\sum_{n=1}^{3}\sum_{m=1}^{5} \dfrac{a_{m+n+k} (1 + Sin(X_1^m) + Sin(X_2^n) + Sin(X_3^k))}{b_{m+n+k} (1 + Sin(X_1^m) + Sin(X_2^n) + Sin(X_3^k))}$	0.993	1.01	0.942

Table 10.5 Coefficient list of models for tension line.

Coefficients	Model 1*	Model 2	Model 3*	Model 4*
a_1	2934	291085.1	129.3	297.5
a_2	36.70	3873.3	229.3	3.181
a_3	−3.0731	68.178	45.56	−1.196
a_4	−24.11	−1944.5	85.52	3.512
a_5	−34.78	−2964.9	−151.9	−4.686
a_6	−40.13	−3476.8	1984	5.278
a_7	480.4	39608.3	1893	44.26
a_8	482.1	42028.8	1264	40.04
a_9	354.1	30321.0	−4280	23.23
a_{10}	−952.8	14210.1	−3629	158.3
a_{11}	−799.3	−60570.9	834.1	−36.72
a_{12}	210.6	−65212.2	–	–
a_{13}	–	−24185.5	–	–
a_{14}	–	30679.2	–	–
b_1	5104	501044	21660	499.7
b_2	74.64	7396.86	510.1	6.503
b_3	12.38	1320.83	106.7	0.10951
b_4	−20.57	1893.11	−106.6	−3.607
b_5	37.27	−3522.37	−214.7	−5.380
b_6	−45.65	−4339.79	−295.7	−6.270
b_7	−72.17	−9633.18	−1426	−5.548
b_8	−345.3	−24438.68	−1671	−29.93
b_9	−411.7	−24101.50	1581	−33.81
b_{10}	367.7	−17530.59	−2014	15.23
b_{11}	−448.2	30061.77	−7187	−250.9
b_{12}	−1616	−2948.49	–	–
b_{13}	–	−55373.26	–	–
b_{14}	–	−109342.53	–	–

* Coefficients of model 1, model 3 and model 4 are divided by (10^7).

3) **Optimization:** This engineering design problem is investigated by introducing eight optimization problems which include minimizing nacelle acceleration and line tension with appropriate constraints (see Table 10.10). As shown in Table 10.10, for the first and second problems, nacelle acceleration and line tension in continuous domains are minimized based on the DE algorithm. In problem 3, four

Table 10.6 Models for nacelle acceleration.

Model No.	Model	R^2 Training	R^2 Training (Adjusted)	R^2 Testing
1	$\sum_{n=1}^{10} \dfrac{a_n (1 + X_1 + X_2 + X_3)^2}{b_n (1 + X_1 + X_2 + X_3)^2}$	1.00	1.00	0.954

Table 10.7 Coefficients list of models for nacelle acceleration.

Coefficients	Model	Coefficients	Model
a_1	3.799	b_1	6.158
a_2	7.362	b_2	19.253
a_3	1.611	b_3	4.890
a_4	0.595	b_4	−15.925
a_5	7.396	b_5	1.386
a_6	5.867	b_6	−2.290
a_7	4.946	b_7	−1.745
a_8	8.129	b_8	23.945
a_9	10.644	b_9	21.348
a_{10}	−14.403	b_{10}	−17.933

Table 10.8 Maximum and minimum values for nacelle acceleration.

Model	Max	Min
1	1.05	0.5

Table 10.9 Maximum and minimum values for line tension.

Models	Max	Min
1	0.915	0.534
2	0.918	0.546
3	0.904	0.551
4	0.914	0.547

optimization algorithms (DE, RS, SA, NM) in discrete domains are used with equality constraints to minimize the line tension.

Lee et al. [14] constructed a framework for optimal design based on the Neuro-Response Surface Method (NRSM) and confirmed the usefulness of the constructed framework using MATLAB software for hydrodynamics performance. The design alternatives for

Table 10.10 Optimization problems.

Problem No.	Objectives	Optimization Algorithms	Constraints
1	Minimizing NA	DE	$0.5 \leq S \leq 1$ $0.5 \leq R \leq 1$ $0.5 \leq H \leq 1$
2	Minimizing LT	DE	$0.5 \leq S \leq 1$ $0.5 \leq R \leq 1$ $0.5 \leq H \leq 1$
3	Minimizing LT	DE, RS, SA, NM	R = 0.5 or R = 1 S = 0.5 or S = 0.75 or S= 1 H = 0.5 or H = 0.75 or H = 1
4	Minimizing LT	DE	R = 0.5 or R = 1 S = 0.5 or S = 0.75 or S = 1 H = 0.5 or H = 0.75 or H = 1 NA < 0.5049
5	Minimizing LT	DE	$0.5 \leq S \leq 1$ $0.5 \leq R \leq 1$ $0.5 \leq H \leq 1$ NA < 0.5049
6	Minimizing LT	DE	$0.5 \leq S \leq 1$ $0.5 \leq R \leq 1$ $0.5 \leq H \leq 1$ NA < 0.5001
7	Minimizing LT	DE	R = 0.5 or R = 1 S = 0.5 or S = 0.75 or S = 1 H = 0.5 or H = 0.75 or H = 1 NA < 0.5001
8	Minimizing NA	DE	$0.5 \leq S \leq 1$ $0.5 \leq R \leq 1$ $0.5 \leq H \leq 1$ LT < 0.5588

NA: Nacelle Acceleration, LT: Line Tension, S: The submerged depth, R: The radius of the cylinder, H: The height of the cylinder.

performance analysis were created by using an orthogonal array table through commercial codes. According to the Pareto-Optimum Set, the results of nacelle acceleration (NA) and line tension (LT) are 0.5049 and 0.5588, respectively [13, 25]. Therefore, in problems 4 and 5, line tension in discrete and continuous domains are minimized with the inequality constraint NA < 0.5049. In problems 6 and 7, to provide Pareto-Optimum Set values of LT, it is minimized with the inequality constraint NA < 0.5001 in discrete and continuous domains. In problem 8, Nacelle acceleration is minimized in a continuous domain with inequality constraint LT < 0.5588 to provide Pareto-Optimum Set values of NA.

Results and Discussion

In this section, results of FOWT's optimal design studies based on hybrid neuro-regression and stochastic optimization are presented for NA and LT. Tables 10.11 and 10.12 show the accuracy of models for training and test data. In these tables, "reference values" are obtained from the commercial code (AQWA) analysis results [14], and "predicted values" are relevant to the output of the neuro-regression analysis in Mathematica. It is observed that with the constructed framework, the majority of the error values are below 0.05 (see Table 10.13).

In this study, to validate the construction of the objective function for optimization, the data provided by Lee et al. [14] is used, and the results are compared with their results. It is also found that neuro-regression analysis gives reasonable results.

As mentioned above, the main objective of the present chapter is to minimize NA and LT by considering the geometry of the TLP-type wind turbine substructure. In this regard, hydrodynamic parameters (NA and LT) are optimized with the appropriate constraints for eight different optimization problems (see Table 10.14).

With these problems, an attempt was made to find values close to the optimum values (NA = 0.5049, LT = 0.5558) obtained from Pareto

Table 10.11 Accuracy of model for training data.

Case	Reference Values [14]		Predicted Values	
	Nacelle Acceleration	Line Tension	Nacelle Acceleration	Line Tension
1	0.780	0.500	0.780	0.57033
2	0.590	0.559	0.590	0.55959
4	0.551	0.555	0.551	0.54802
5	0.670	0.586	0.670	0.57974
6	0.500	0.576	0.500	0.56381
7	0.913	0.640	0.913	0.66231
9	0.832	0.660	0.832	0.67087
10	0.607	0.695	0.607	0.66018
11	0.754	0.714	0.754	0.72131
12	0.553	0.773	0.553	0.70941
14	0.714	0.798	0.714	0.77349
15	0.899	0.688	0.899	0.81469
16	0.649	0.706	0.649	0.80758
17	0.790	1.000	0.790	0.91416
18	0.589	0.940	0.589	0.90883

Table 10.12 Accuracy of the model for the test data.

Case	Reference Values [14]		Prediction Values	
	Nacelle Acceleration	Line Tension	Nacelle Acceleration	Line Tension
3	0.723	0.587	0.736807	0.56109
8	0.665	0.667	0.656895	0.65324
13	1.000	0.770	1.05152	0.78018

Table 10.13 Error prediction for the test data.

Case	Error [(Prediction value – Reference Value)/Reference Value]	
	Nacelle Acceleration	Line Tension
3	0.019	0.04
8	0.012	0.02
13	0.05	0.01

Table 10.14 Optimum results.

Problem No.	Optimum Values	Design Variables		
		R	S	H
1	NA = 0.5	1	0.5	1
2	LT = 0.5477	1	0.5	0.7164
3*	LT = 0.5480	1	0.5	0.75
4	LT = 0.5638	1	0.5	1
5	LT = 0.5620	1	0.5	0.9791
6	LT = 0.5637	1	0.5	0.9995
7	LT = 0.5638	1	0.5	1
8	NA = 0.5135	1	0.5	0.8125

* Different stochastic methods and Nelder–Mead method are used for the third problem. It is found that each method provides the same optimum value.

Optimum Set given in the referenced article. Design variables "R", "S", "H" are also calculated for each problem. A close look at the results given in Table 10.14 demonstrates that in this study, optimum values of nacelle acceleration and line tension are close to the optimum values which have been obtained by Pareto Optimum Set in reference article.

Conclusion

In the present study, the hydrodynamics effect on the substructure of NREL 5 MW TLP-type FWOT are investigated. Multiple nonlinear regression analysis models using Mathematica to improve the hydrodynamic performance of the NREL 5 MW TLP-type FWOT are employed. According to these models, hydrodynamic performance parameters are analyzed in the continuous-discrete and discrete domains to minimize the outputs using stochastic optimization methods which are DE, RS, SA and a deterministic one Nelder-Mead. As a result, lower error values are achieved with the proposed framework. The present study reveals the importance of input parameters and their effect on the nacelle acceleration and line tension. Besides that, these methods can be used for the effectiveness and usefulness of various geometry optimization problems in ocean engineering and naval architecture.

References

[1] Rueda-Bayona, J.G., Guzmán, A., Eras, J.J.C., Silva-Casarín, R., Bastidas Arteaga, E. and Horrillo-Caraballo, J. 2019. Renewables energies in Colombia and the opportunity for the offshore wind technology. J. Clean. Prod.

[2] Elkinton, Christopher N., James F. Manwell and Jon G. McGowan. 2008. Optimizing the layout of offshore wind energy systems. Mar. Technol. Soc. J. 42.2: 19–27.

[3] Barthelmie, R.J. 1993. Prospects for offshore wind energy: The state of the art and future opportunities: Offshore wind climate and wind energy applications. J. Wind Eng. (1993): 86–99.

[4] Http://offshorewindenergy.org , (2008), March.

[5] Namik, H. and Stol, K. 2011. Performance analysis of individual blade pitch control of offshore wind turbines on two floating platforms. 21: 691–70.

[6] Kurian, V.J., Narayanan, S.P. and Ganapathy, C. 2010. Towers for Offshore Wind Turbines. The 10th Asian International Conference on Fluid Machinery. Malaysi

[7] Bui, C., Nie, R., Marini, S., Ibrovic, V. and Knijnenbur, L. 2019, April. Http://floatingwindfarm.weebly.com . Floating Wind Turbines.

[8] Li, L., Liu, Y., Yuan, Z. and Gao, Y. 2019. Dynamic and structural performances of offshore floating wind turbines in turbulent wind flow. Ocean Eng. 179: 92–103.

[9] NREL Transforming Energy. https://www.nrel.gov/.

[10] Oguz, E., Clelland, D., Day, A.H., İncecik, A., Lopez, J.A., Sánchez, G. and Almeria, G.G. 2018. Experimental and numerical analysis of a TLP floating offshore wind turbine. Ocean Eng. 147: 591–605.

[11] Martin, H.R., Kimball, R.W., Viselli, A.M. and Goupee, A.J. 2014. Methodology for wind/wave basin testing of floating offshore wind turbines. J. Offshore Mech. Arct. 136.

[12] Bachynski, E.E. and Moan, T. 2012. Design considerations for tension leg platform wind turbines. Mar. str. 29: 89–114.

[13] Zamora-Rodriguez, R., Gomez-Alonso, P., Amate-Lopez, J., De-Diego-Martin, V., Dinol, P., Simos, A.N. and Souto-Iglesias, A. 2014. Model scale analysis of a TLP floating offshore wind turbine. Offshore and Arctic Engineering. Proceedings of the ASME 2014 33rd International Conference on Ocean, Offshore and Arctic Engineering, OMAE2014, San Francisco, California, USA.

[14] Lee, J.J., Shin, S.C. and Kim, S.Y. 2015. An optimal design of wind turbine and ship structure based on neuro-response surface method. Int. J. Nav. Archit. Ocean Eng. 7: 750–769.

[15] Ramachandran, K. and Vasanta, G.K. 2013. A Numerical Model for a Floating TLP Wind Turbine. M.S. Thesis, Technical University of Denmark, Denmark.

[16] Karem, R.A., Aktas, K., Ozbahceci, B. and Ozkol, U. 2019. Açık deniz rüzgar türbinleri yüzer platformlarının hidrodinamik modellemesi, 9. Kıyı mühendisliği sempozyumu, Ankara, 788–801.

[17] Yang, H., Zhu, Y., Lu, Q. and Zhang, J. 2015. Dynamic reliability based design optimization of the tripod sub-structure of offshore wind turbines. Renew. Energy 78: 16–25.

[18] Damiani, R.R. 2016. Design of offshore wind turbine towers. Arvada, United States. pp. 263–357.

[19] Yang, J., He, E.M. and Hu, Y.Q. 2019. Dynamic modelling and vibration suppression for an offshore wind turbine with a tuned mass damper in floating platform. Appl. Ocean Res. 83: 21–29.

[20] Savran, M. and Aydin, L. 2018. Stochastic optimization of graphite-flax/epoxy hybrid laminated composite for maximum fundamental frequency and minimum cost. Eng. Struct. 174: 675–687.

[21] Sreenivas, P. and Kumar, S.V. 2015. A review on non-traditional optimization algorithm for simultaneous scheduling problems. J. Mech. Civ. Eng. 12: 50–53, India.

[22] Aydin, L. and Artem, H.S. 2017. Design and Optimization of Fiber Composites. Woodhead Publishing Series in Composites Science and Engineering, Turkey, pp. 299–314.

[23] Gentils, T., Wang, L. and Kolios, A. 2017. Integrated structural optimisation of offshore wind turbine support structures based on finite element analysis and genetic algorithm. Appl. Energy 199: 187–204.

[24] Rao, P.V., Aravindan, S. and Bhavsar, S.N. 2014. Investigating material removal rate and surface roughness using multi-objective optimization for focused ion beam (fib) micro-milling of cemented carbide. Precision Engineering 40: 131–138.

[25] Lee, J.C., Shin, S.C. and Kim. 2013. Development of framework for NRSM based optimal shape design. Proceedings of KIIS Spring Conference, Korean.

CHAPTER 11

Structural Optimization of Bulk Carrier Bottom Stiffened Panels by Using Multiple Nonlinear Regression Model

Mehmet Sari,[1] *İlhan Tatar,*[2] *Levent Aydin,*[2]
Selda Oterkus[3],* and *Erkan Oterkus*[3]

Introduction

Optimization methods are of great importance in structural design; they allows a designer or decision-maker to find the best solutions from the resources available. Structural optimization has found a wide range of applications by skilfully combining mathematical and mechanical knowledge with engineering, which has attracted the attention of many researchers. For example, in [1], which characterizes the main load effects in ship structures, a final formulation aimed at estimating wave-induced load effects is presented by making use of the stagnant water load effect. They concluded that a nonlinear analysis is needed for wave-induced load estimation, but they linearly predict the main effects. However, they stated that the critical effect of stagnant water load is fatigue. In [2], the

---

[1] Volcano Fire Materials Co., Ltd. Torbalı/İzmir.
   Email: m.sari@outlook.com.tr
[2] İzmir Katip Çelebi University, Faculty of Engineering and Architecture, Department of Mechanical Engineering.
   Emails: ilhan086@gmail.com; leventaydinn@gmail.com
[3] University of Strathclyde, PeriDynamics Research Centre, Department of Naval Architecture, Ocean & Marine Engineering, Glasgow, Scotland.
   Email: erkan.oterkus@strath.ac.uk
* Corresponding author: selda.oterkus@strath.ac.uk

researchers have developed a computer-aided method for the optimal design of stiffened panels, which is the main structural component of a ship under reliability constraints. The problem is formulated according to a particular structural configuration and loading situation. As a result of the study, it was determined that approximately 15–20% by weight could be saved without reducing the safety of the stiffened panel. A generalized method is examined for the design of ship hulls by FEM results of stress distribution and damage mechanisms [3]. It was concluded that the stress distribution depends on the initial defects. It also stated that the numerical predictions can be made for trunk beam exposed to buckling by a linear approach and Smith method; however, it is necessary to change the approach in compound loading conditions. A ship's structure is examined statically and dynamically, and a multi-purpose problem is presented in [4]. The aim is to investigate the multi-objective optimization problems of structural elements used in ship structures based on genetic algorithm. The results showed that the design of ship structural parts could be made successfully with the proposed design process. In another study on design optimization of a ship structure, a collaborative optimization model has been considered [5]. The main objective is to provide weight saving, increasing strength, and reducing the vibration level. It concluded that the proposed model can be applied to solve the real-life problems of naval engineering. Mathematical optimization of a ship structure has been investigated based on the FEM in the structural evaluation process in [6]. The most significant contribution of the study is that it performs design, finite element analysis, and optimization process as an integrated software. The aim of the study was to evaluate the reliability of the ship's structures by using nonlinear finite element analysis results. It was concluded that the sample amount was not significant for the First Order Reliability Method (FORM) while it was important for Maximum Entropy Fitting Method (MEM). Besides, the FORM was suitable for probabilistic modeling [7]. Taylor series expansion method and Monte Carlo simulation were used, and the ultimate strength of the ship's structure exposed to corrosion was evaluated statistically in [8]. The effect of the variables on the target function was evaluated by correlation analysis, and the probability density function was also investigated. It was concluded that the trunk beam modulus coefficient of variation and ultimate bending strength values was significantly changed by the variables. In [9], a multi-objective optimization model based on genetic algorithm has been established in order to minimize weight and total surface area for a ship's structure. The simulation results for a high-speed catamaran were evaluated statistically, and the optimization equation was generated. A hatch cover of a bulk carrier has been optimized, and the procedure was used to evaluate the optimum dimensions of the structure. Sequential Quadratic Programming

(SQP) based optimization program has been developed, and 8.5% weight reduction has been provided in the ship structure [10]. Recently, the effects of the buckling capacities of the stiffened panels on the geometric degradation of two main distortions, local and global, were evaluated [11]. Local disturbances were examined under three headings: plate, web, and stiffener. As a result of the study, they stated that only local plate and web crashes are effective in cases where maximum global bending stress is effective, and stiffener and local web degradations should be included in cases where global degradation is maximum. Dimensional and simulation analyses of stiffened panels exposed to the buckling effect were performed, and the damage behaviour of ship structures was investigated with small scale models in [12]. As a result of the study, it has been found that simulations of ultimate strength and damage modes are consistent. Optimization of the conveyance and measure of the stiffeners at the same time by creating a hereditary calculation with a genetic algorithm with two-level approximation (GATA) has been performed in [13]. A strategy for truss format optimization was introduced utilizing GATA. Numerical illustrations appear that progressed GATA can optimize the topology and measure factors of the solidified shell with generally high productivity simultaneously. As a result of the optimization, the structure weight was diminished by 10.66%. For a passenger ship, finite element modeling, developmental optimization calculation, and particle swarm optimization (PSO) calculation utilizing circuitous limitation releasing have been considered in [14]. It has been concluded that the usage of ESL components is reasonable for the optimization of expensive, complex structures such as cutting-edge passenger ships.

The purpose of the present study is to determine the dimensions of the bottom stiffened panels of the bulk carrier, taking into account the structural performances by considering ultimate strength and weight. The optimization objective of the design of the bottom plate of a bulk carrier is chosen with the need for structural safety and weight reduction. The structural geometry is optimized for a minimum weight and maximum strength by using the nonlinear global optimization method considering general and industrial constraints. The optimized dimensions of the stiffened panel are obtained by using the neural regression method and objective function.

Methodology

The most critical design criteria affecting ship structural design can be listed as weight, strength, reliability, and production processes of the structure. Stiffened steel panels are widely used as the primary carrier system element in maintaining the structural integrity of the ships. The development of a methodology that decreases the design process time of these intricate structures has been one of the research interests of designers. This study aims to enhance the structural performance (steel weight and ultimate strength) by optimizing the design variables of stiffened panels. The properties of the structural element examined in the reference study are given in Table 11.1. Five design parameters of stiffened panel geometry are selected for the optimization problem [15]. These parameters are provided in Table 11.2. Figure 11.2 represents the design variables.

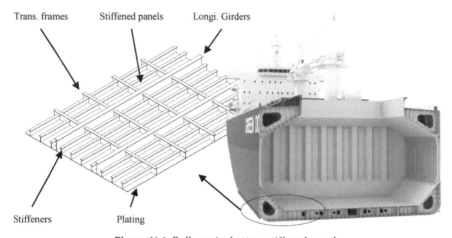

Trans. frames Stiffened panels Longi. Girders

Stiffeners Plating

Figure 11.1 Bulk carrier bottom stiffened panels.

Table 11.1 Properties of bottom stiffened panels of a bulk carrier [15].

Definition	Symbol	Values
Yield stress	σ_{yp}	313.6 N/mm²
Elastic modulus	E	205800 N/mm²
Poisson's ratio	N	0.3
Plate length	A	2550 mm
Plate breath	B	850 mm

Table 11.2 Design parameters and their ranges [15].

Symbol	Definition	Range of Design Parameters
t_p (mm)	Plate thickness	{9.5,13,16}
t_w (mm)	Web thickness	{9,10,12,15,17}
t_f (mm)	Flange thickness	{12,15,17,19,20}
h_w (mm)	Web height	{138,150,235,383,580}
b_f (mm)	Flange width	{70,80,90,130,150}

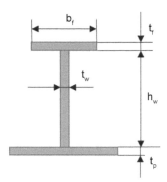

Figure 11.2 Cross-section of the stiffened panel.

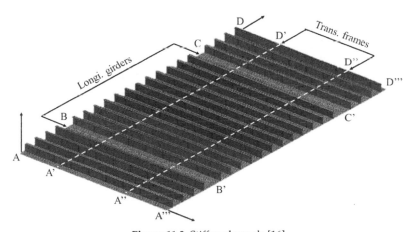

Figure 11.3 Stiffened panels [16].

Figure 11.3 represents the stiffened panel considered in the reference study [15]. The structural analysis results are presented in Table 11.3. In the reference study, all data were normalized between the value of 0.5 to 1 to increase the learning rate of a neural network, as shown in Table 11.4.

Multiple Nonlinear Regression Model

Regression analysis is used to estimate the relationship between parameters affecting engineering events. Multiple regression allows you to determine the overall suitability of the model and the relative contribution of each estimator to the total variance described. Regression models include linear, nonlinear, rational, and logistic, etc. In general, engineering systems involve nonlinear behaviours such as logarithmic, exponential, and trigonometric. Before the optimization process, the objective function, design variables, and constraints of the optimization problem must be defined.

Table 11.3 Design alternatives and results of structural analysis [15].

Case	t_p (mm)	t_w (mm)	t_f (mm)	h_w (mm)	b_f (mm)	Steel Weight (kg)	Ultimate Strength (MPa)
1	9.5	9.0	12.0	138.0	70.0	1797.295	0.531
2	9.5	10.0	15.0	150.0	80.0	1896.766	0.520
3	9.5	12.0	17.0	235.0	90.0	2162.343	0.575
4	9.5	15.0	19.0	383.0	130.0	2784.438	0.714
5	9.5	17.0	20.0	580.0	150.0	3532.079	0.906
6	13.0	9.0	15.0	235.0	130.0	2655.170	0.428
7	13.0	10.0	17.0	383.0	150.0	3027.784	0.499
8	13.0	12.0	19.0	580.0	70.0	3335.210	0.651
9	13.0	15.0	20.0	138.0	80.0	2591.593	0.739
10	13.0	17.0	12.0	150.0	90.0	2585.155	0.719
11	16.0	9.0	17.0	580.0	80.0	3521.717	0.497
12	16.0	10.0	19.0	138.0	90.0	2959.981	0.969
13	16.0	12.0	20.0	150.0	130.0	3170.833	1.122
14	16.0	15.0	12.0	235.0	150.0	3319.718	1.229
15	16.0	17.0	15.0	383.0	70.0	3679.615	1.296

Table 11.4 Training data [15].

Case	t_p (mm)	t_w (mm)	t_f (mm)	h_w (mm)	b_f (mm)	Steel Weight (kg)	Ultimate Strength (MPa)
1	0.500	0.500	0.500	0.500	0.500	0.500	0.856
2	0.500	0.563	0.688	0.514	0.563	0.526	0.868
3	0.500	0.688	0.813	0.610	0.625	0.597	0.809
4	0.500	0.875	0.938	0.777	0.875	0.762	0.701
5	0.500	1.000	1.000	1.000	1.000	0.961	0.606
6	0.769	0.500	0.688	0.610	0.875	0.728	1.000
7	0.769	0.563	0.813	0.777	1.000	0.827	0.894
8	0.769	0.688	0.938	1.000	0.500	0.909	0.744
9	0.769	0.875	1.000	0.500	0.563	0.711	0.686
10	0.769	1.000	0.500	0.514	0.625	0.709	0.698
11	1.000	0.500	0.813	1.000	0.563	0.958	0.896
12	1.000	0.563	0.938	0.500	0.625	0.809	0.583
13	1.000	0.688	1.000	0.514	0.875	0.865	0.538
14	1.000	0.875	0.500	0.610	1.000	0.904	0.513
15	1.000	1.000	0.688	0.777	0.500	1.000	0.500

In this chapter, we introduce a novel approach to the modeling-design-optimization process in order to optimize the engineering input parameters. Firstly, a detailed study on multiple nonlinear neuro-regression analysis, including linear, quadratic, trigonometric, logarithmic, and their rational forms for the prescribed problem (output) are performed (A detailed explanation on multiple nonlinear neuro-regression analysis is given in Chapter 1). Secondly, boundednesses of the candidate models are checked to provide generating realistic values. Finally, the different direct search methods, including stochastic ones, are performed. The equation used for the polynomial model is given as

$$\sum_{k=0}^{n} \sum_{p=0}^{(n-k)} \sum_{q=0}^{(n-k-p)} \sum_{s=0}^{(n-k-p-q)} \sum_{r=0}^{(n-k-p-q-s)}$$

$$\left(\frac{n!}{(n-k-p-q-s-r)!\,k!\,p!\,q!\,s!\,r!} \right) x_5^k \, x_4^p \, x_3^q \, x_2^s \, x_1^r \, 1^{(n-k-p-q-s)} \tag{11.1}$$

Table 11.5 represents the functions used for regression models.

Table 11.5 Functions for the regression model.

Model Name	Symbol	Definition
Polynominal func.	P_n	n^{th} ordered polynomial function
Polynomial rational func.	P_{nr}	n^{th} ordered, rational polynomial function
Logarithmic func.	L_n	n^{th} ordered logarithmic function
Logarithmic, rational func.	L_{nr}	the nth ordered, rational logarithmic function
Trigonometric func.	T_n	n^{th} ordered trigonometric function
Trigonometric rational func.	T_{nr}	n^{th} ordered, rational, trigonometric function

Determination of the Multiple Nonlinear Neuro-Regression Models

In this study, the artificial neural network approach is combined with a multi-stage regression model. For this hybrid approach, the numerical data is divided into two random sections, eighty percent of the data is selected as training data, and twenty percent of the data is selected as test data. ANN approach is applied to the regression model with randomly selected test data.

Results and Discussions

In this study, statistical analysis is carried out based on a reference study [15] by using the results of the finite element analysis for five independent variables. The objective function is selected by examining R^2 values of thirty different functions by using multiple nonlinear regression and ANN hybrid model. The best-fitting correlation is obtained by the multiple stepwise regression model and the results provided in Tables 11.6 and 11.7.

According to the objective function in Table 11.6, the regression model of steel weight (y_1) of the stiffened panel can be expressed as:

$$y_1 = (16.0777 + 140.043x_1 + 5.866x_1^2 - 65.0629x_2 - 3.78021x_1x_2 + 0.60611x_2^2 - 0.86344x_3 + 6.06658x_1x_3 - 14.8751x_2x_3 - 1.03607x_3^2 - 144.163x_4 - 10.3046x_1x_4 + 23.291x_2x_4 + 7.38265x_3x_4 + 1.19627x_4^2 + 19.3953x_5 - 3.39725x_1x_5 + 0.427512x_2x_5 + 13.1186x_3x_5 + 11.5711x_4x_5 - 0.171023x_5^2)/(336.018 - 9.17065x_1 - 1.99729x_1^2 - 129.579x_2 + 2.15831x_1x_2 - 1.46134x_2^2 - 23.7299x_3 - 1.73321x_1 x_3 + 6.85739x_2x_3 + 1.69939x_3^2 + 57.8343x_4 + 17.3509x_1x_4 - 19.782x_2x_4 - 1.04478x_3x_4 + 1.81908x_4^2 + 9.96606x_5 + 9.06684x_1x_5 - 1.73137x_2x_5 - 10.3282x_3x_5 - 6.31225x_4 x_5 + 1.40766x_5^2)$$

$$(11.2)$$

According to objective function in Table 11.7, the regression model of ultimate strength (y_2) of stiffened panel can be expressed as: (The original (long) form of the output function is given in Appendix)

$$y_2 = (0.573443 + 14.4466 \cos x_1 + \ldots + 0.220581 \sin x_5^3)/(1.67674 + 8.03971 \cos x_1 + \ldots + 1.08223 \sin x_5^3)$$

$$(11.3)$$

Boundedness Check for the Models

To perform the boundedness check of the selected models, the values of the five independent variables are increased and decreased by 1%, 5%, 10%, 25%, 50%. The main goal of this step is to investigate the variation of the results. If the generated values stay in the realistic range in terms of engineering, it means that the candidate model is appropriate. Tables 11.8 and 11.9 present the data obtained as a result of the sensitivity for the outputs steel weight and ultimate strength, respectively. It is observed that both of the models for y_1 and y_2 are also appropriate. In another word, all the generated values given in the tables are also in the realistic range.

Optimization of Stiffened Panel for Ship Structure

In the optimization step, it is aimed to lighten the weight and increase the ultimate strength of the stiffened panel structures, which are the crucial elements of the ship's structural design. For this purpose, four different optimization problems are introduced as provided in Table 11.10.

Table 11.6 Results of regression modeling of steel weight (y_1) for stiffened panel.

No.	Function Symbol	R^2	R^2_{adjust}	$R^2_{testing}$
1	P_1	0.999384	–	0.967946
2	P_2	1	1	0.988048
3	P_3	1	1	0.965818
4	P_4	1	1	0.956784
5	P_5	1	1	0.966236
6	P_{1r}	0.999986	-	0.892812
7	P_{2r}	1	1	0.993414
8	P_{3r}	1	1	0.985687
9	P_{4r}	1	1	0.978003
10	P_{5r}	1	1	0.970437
11	L_1	0.998965	–	0.888663
12	L_2	1	1	–1.01956
13	L_3	1	1	–4.20757
14	L_4	1	1	–8.10035
15	L_5	1	1	–12.439
16	L_{1r}	0.999976	–	–4.5261
17	L_{2r}	1	1	0.880826
18	L_{3r}	1	1	0.666194
19	L_{4r}	1	1	0.484543
20	L_{5r}	1	1	–0.428246
21	T_1	1	1	0.956452
22	T_2	1	1	0.99491
23	T_3	1	1	0.99445
24	T_4	1	1	0.993728
25	T_5	1	1	0.993089
26	T_{1r}	1	1	0.99901
27	T_{2r}	1	1	0.995609
28	T_{3r}	1	1	0.995584
29	T_{4r}	1	1	0.995652
30	T_{5r}	1	1	0.99636

Table 11.7 Results of regression modeling of ultimate strength (y_2) for stiffened panel.

No	Function Symbol	R^2	R^2_{adjust}	$R^2_{testing}$
1	P_1	0.989602	–	0.448218
2	P_2	1	1	0.567349
3	P_3	1	1	0.554882
4	P_4	1	1	0.414243
5	P_5	1	1	–0.0207094
6	P_{1r}	0.998147	–	–0.175372
7	P_{2r}	1	1	0.788842
8	P_{3r}	1	1	0.724272
9	P_{4r}	1	1	0.680045
10	P_{5r}	1	1	0.657979
11	L_1	0.989127	–	0.300762
12	L_2	1	1	0.714861
13	L_3	1	1	–0.461564
14	L_4	1	1	–3.3729
15	L_5	1	1	–9.75379
16	L_{1r}	0.999987	–	0.593852
17	L_{2r}	1	1	–0170547
18	L_{3r}	1	1	–8027.74
19	L_{4r}	1	1	–1205
20	L_{5r}	1	1	–704.29
21	T_1	0.997304	1.00324	0.0120144
22	T_2	1	1	0.883604
23	T_3	1	1	0.859305
24	T_4	1	1	0.854328
25	T_5	1	1	0.85495
26	T_{1r}	1	1	–2.61869
27	T_{2r}	1	1	0.912242
28	T_{3r}	1	1	0.913358
29	T_{4r}	1	1	0.913381
30	T_{5r}	1	1	0.912986

Table 11.8 Boundedness check of weight model (y_1).

–50%	–25%	–20%	–15%	–10%	–5%	–1%	Exp.	1%	5%	10%	15%	20%	25%	50%
0.25	0.37	0.39	0.42	0.45	0.47	0.49	0.50	0.51	0.53	0.56	0.59	0.62	0.65	0.82
0.25	0.38	0.41	0.44	0.47	0.50	0.52	0.53	0.53	0.56	0.59	0.62	0.66	0.70	0.89
0.27	0.42	0.45	0.49	0.52	0.56	0.59	0.60	0.60	0.64	0.68	0.72	0.76	0.81	1.08
0.32	0.51	0.56	0.60	0.65	0.71	0.75	0.76	0.77	0.82	0.88	0.95	1.02	1.09	1.55
0.38	0.63	0.69	0.75	0.81	0.89	0.95	0.96	0.98	1.04	1.13	1.22	1.32	1.43	2.13
0.35	0.53	0.57	0.61	0.65	0.69	0.72	0.73	0.74	0.77	0.81	0.86	0.90	0.95	1.20
0.39	0.60	0.65	0.69	0.74	0.79	0.83	0.83	0.85	0.89	0.94	0.99	1.05	1.10	1.41
0.42	0.65	0.70	0.75	0.80	0.85	0.90	0.91	0.92	0.97	1.02	1.08	1.14	1.21	1.55
0.30	0.48	0.52	0.57	0.61	0.66	0.70	0.71	0.72	0.76	0.82	0.88	0.94	1.01	1.42
0.30	0.48	0.52	0.56	0.61	0.66	0.70	0.71	0.72	0.76	0.82	0.88	0.95	1.02	1.46
0.47	0.71	0.76	0.81	0.86	0.91	0.95	0.96	0.97	1.01	1.06	1.11	1.17	1.22	1.50
0.38	0.58	0.62	0.67	0.71	0.76	0.80	0.81	0.82	0.86	0.91	0.97	1.02	1.08	1.39
0.39	0.61	0.65	0.71	0.76	0.81	0.86	0.87	0.88	0.93	0.99	1.05	1.12	1.19	1.58
0.40	0.63	0.68	0.73	0.79	0.84	0.89	0.90	0.92	0.97	1.03	1.10	1.17	1.24	1.66
0.42	0.68	0.73	0.80	0.86	0.93	0.99	1.00	1.01	1.08	1.15	1.24	1.33	1.42	1.98

Table 11.9 Boundedness check of ultimate strength model (y_2).

–50%	–25%	–20%	–15%	–10%	–5%	–1%	Exp.	1%	5%	10%	15%	20%	25%	50%
0.77	0.87	0.88	0.89	0.90	0.90	0.90	0.86	0.90	0.90	0.90	0.89	0.88	0.86	0.75
0.79	0.87	0.88	0.88	0.88	0.87	0.87	0.87	0.87	0.86	0.84	0.83	0.81	0.79	0.63
0.80	0.85	0.85	0.85	0.84	0.82	0.81	0.81	0.81	0.79	0.77	0.74	0.72	0.68	0.50
0.83	0.83	0.81	0.79	0.76	0.73	0.71	0.70	0.69	0.67	0.63	0.59	0.54	0.50	0.25
0.83	0.78	0.76	0.72	0.69	0.65	0.61	0.61	0.60	0.56	0.51	0.46	0.41	0.36	0.10
0.95	1.07	1.07	1.07	1.05	1.03	1.01	1.00	0.99	0.96	0.90	0.84	0.77	0.69	0.16
0.99	1.09	1.08	1.06	1.04	1.00	0.96	0.89	0.94	0.89	0.82	0.74	0.65	0.55	–0.04
0.85	0.87	0.86	0.84	0.81	0.78	0.75	0.74	0.74	0.70	0.65	0.60	0.54	0.48	0.12
0.81	0.81	0.80	0.78	0.75	0.72	0.69	0.69	0.68	0.65	0.61	0.56	0.51	0.46	0.19
0.81	0.81	0.80	0.78	0.76	0.73	0.70	0.70	0.69	0.66	0.62	0.58	0.54	0.49	0.22
0.96	1.05	1.05	1.03	1.00	0.95	0.91	0.90	0.88	0.82	0.73	0.62	0.50	0.35	–0.53
0.87	0.85	0.82	0.77	0.72	0.66	0.60	0.58	0.57	0.50	0.41	0.32	0.22	0.12	–0.41
0.91	0.85	0.80	0.74	0.67	0.60	0.53	0.54	0.49	0.42	0.32	0.21	0.10	–0.01	–0.53
0.92	0.85	0.81	0.75	0.68	0.60	0.53	0.51	0.49	0.42	0.32	0.21	0.10	–0.01	–0.54
0.81	0.74	0.70	0.66	0.61	0.56	0.51	0.50	0.49	0.44	0.37	0.30	0.23	0.15	–0.23

The objective functions obtained with the multiple neuro-regression models is optimized by taking into account the constraints provided in Table 11.10. For this purpose, the Mathematica program is used. The optimization study is carried out using the minimum and maximum value range of system variables provided in problems 1 and 2. The aim is to examine the limits of the model mathematically. Tables 11.11–11.14 show the optimization results of the problems 1–4. Table 11.11 gives the optimum values of the parameters t_p, t_w, t_f, h_w, and bf that minimize y_1. It is also shown that the optimum values of the input parameters t_p, t_w, t_f, h_w, and b_f that minimize y_2 in Table 11.12. In problem 1, the weight of the structure is aimed to be minimum, provided that the final strength value is the maximum. The structural weight is found as 3128.44 kg for the maximum value of the ultimate strength. Compared with the results of the reference study [15], it is observed that the steel weight of the stiffened panel is reduced by 14.98%.

In the second optimization problem, the maximum of the ultimate strength value is examined when the steel weight (y_1) is less than 1896.766. The strength value is found as 0.81 Mpa provided that structural weight

Table 11.10 Four optimization scenarios.

Problem No.	Objectives	Constraints
1	$\min y_1 (t_p, t_w, t_f, h_w, b_f)$	$9.5 \le t_p \le 16$ $9 \le t_w \le 17$ $12 \le t_f \le 20$ $138 \le h_w \le 580$ $70 \le b_f \le 150$ $y_2 \le 1.296$
2	$\max y_2 (t_p, t_w, t_f, h_w, b_f)$	$9.5 \le t_p \le 16$ $9 \le t_w \le 17$ $12 \le t_f \le 20$ $138 \le h_w \le 580$ $70 \le b_f \le 150$ $y_1 \le 1896.766$
3	$\min y_1 (t_p, t_w, t_f, h_w, b_f)$	$t_p \in \{9.5,13,16\}$ $t_w \in \{9,10,12,15,17\}$ $t_f \in \{12,15,17,19,20\}$ $h_w \in \{138,150,235,383,580\}$ $b_f \in \{70,80,90,130,150\}$ $0.428 \le y_2 \le 1.296$
4	$\max y_2 (t_p, t_w, t_f, h_w, b_f)$	$t_p \in \{9.5,13,16\}$ $t_w \in \{9,10,12,15,17\}$ $t_f \in \{12,15,17,19,20\}$ $h_w \in \{138,150,235,383,580\}$ $b_f \in \{70,80,90,130,150\}$ $1797.295 \le y_1 \le 3679.615$

Table 11.11 Optimization results for problem 1.

t_p (mm)	t_w (mm)	t_f (mm)	h_w (mm)	b_f (mm)	Weight (kg)	Ultimate Strength (MPa)
16	15.4	12	138	150	3128.44	1.296

Table 11.12 Optimization results for problem 2.

t_p (mm)	t_w (mm)	t_f (mm)	h_w (mm)	b_f (mm)	Weight (kg)	Ultimate strength (MPa)
9.5	17	12	138	81.32	1896.766	0.81

Table 11.13 Optimization results for problem 3.

t_p (mm)	t_w (mm)	t_f (mm)	h_w (mm)	b_f (mm)	Weight (kg)	Ultimate strength (MPa)
9.5	9	12	138	70	1797.295	0.531

Table 11.14 Optimization results for problem 4.

t_p (mm)	t_w (mm)	t_f (mm)	h_w (mm)	b_f (mm)	Weight (kg)	Ultimate strength (MPa)
16	17	20	138	90	3284.35	1.296

is 1896.766 kg. Compared with the values given in [15], it is observed that when the design parameters of problem 2 are used, the strength value is increased by 35.80%. In the third optimization problem, the steel weight is minimized by using industrial constraints. The values obtained are similar to the reference study.

In problem 4, the maximum value of the ultimate strength (y_2) is examined within the industrial production conditions of the stiffened panel. When the strength value is maximized, the weight of the stiffened panel is calculated as 3284.35 kg. In the reference study, the steel weight value is 3679.615 kg for maximum ultimate strength value. With the proposed design variables, it is shown that it is possible to design a lighter structure with maximum ultimate strength value.

Conclusion

In this study, it was aimed to determine optimal dimensions for the stiffened panel of the bottom part of the bulk carrier by satisfying the ultimate strength and weight constraints. The finite element analysis results are investigated by using multiple neuro-regression approaches. As a result, the stiffened panel is optimized for different scenarios. It is shown that it is possible to design a lighter structure with maximum ultimate strength based on the proposed approach.

For future investigations, the optimal design of stiffened panels for variable cross-sections for different loading conditions can be considered. This methodology can also be used for the design optimization of composite materials.

Appendix

y2=(1.35211   cos^3(x1)+4.09837   cos(x2)   cos^2(x1)+7.88701   cos(x3) cos^2(x1)+15.3309 cos(x4) cos^2(x1)+13.3033 cos(x5) cos^2(x1)+20.519 sin(x1) cos^2(x1)+14.9265 sin(x2) cos^2(x1)+12.7013 sin(x3) cos^2(x1)+2.17399 sin(x4) cos^2(x1)+7.79371   sin(x5)   cos^2(x1)+13.0949   cos^2(x1)+2.76934   cos^2(x2) cos(x1)+6.76639   cos^2(x3)   cos(x1)+18.3372   cos^2(x4)   cos(x1)+13.2693 cos^2(x5) cos(x1)+11.2776 sin^2(x1) cos(x1)+14.6773 sin^2(x2) cos(x1)+10.6802 sin^2(x3) cos(x1)-0.890582 sin^2(x4) cos(x1)+4.1773 sin^2(x5) cos(x1)+16.019 cos(x2) cos(x1)+18.1642 cos(x2) cos(x3) cos(x1)+33.2482 cos(x3) cos(x1)+14.4014 cos(x2)   cos(x4)   cos(x1)+29.8513   cos(x3)   cos(x4)   cos(x1)+56.6565   cos(x4) cos(x1)+4.86902 cos(x2) cos(x5) cos(x1)+26.1984 cos(x3) cos(x5) cos(x1)+67.023 cos(x4)   cos(x5)   cos(x1)+50.9443   cos(x5)   cos(x1)+41.385   cos(x2)   sin(x1) cos(x1)+49.9025 cos(x3) sin(x1) cos(x1)+60.6734 cos(x4) sin(x1) cos(x1)+51.6164 cos(x5)   sin(x1)   cos(x1)+64.3172   sin(x1)   cos(x1)+15.5807   cos(x2)   sin(x2) cos(x1)+26.0746 cos(x3) sin(x2) cos(x1)+68.9012 cos(x4) sin(x2) cos(x1)+70.5137 cos(x5)   sin(x2)   cos(x1)+43.0676   sin(x1)   sin(x2)   cos(x1)+50.9233   sin(x2) cos(x1)+14.1538 cos(x2) sin(x3) cos(x1)+32.7636 cos(x3) sin(x3) cos(x1)+55.3674 cos(x4) sin(x3) cos(x1)+50.4186 cos(x5) sin(x3) cos(x1)+45.8179 sin(x1) sin(x3) cos(x1)+47.262 sin(x2) sin(x3) cos(x1)+46.056 sin(x3) cos(x1)+13.5561 cos(x2) sin(x4)   cos(x1)+17.5137   cos(x3)   sin(x4)   cos(x1)+10.3746   cos(x4)   sin(x4) cos(x1)+3.82116 cos(x5) sin(x4) cos(x1)+25.7197 sin(x1) sin(x4) cos(x1)-13.9058 sin(x2)   sin(x4)   cos(x1)-0.870513   sin(x3)   sin(x4)   cos(x1)+7.21036   sin(x4) cos(x1)+31.2425 cos(x2) sin(x5) cos(x1)+33.7064 cos(x3) sin(x5) cos(x1)+22.3179 cos(x4) sin(x5) cos(x1)+37.3784 cos(x5) sin(x5) cos(x1)+48.1309 sin(x1) sin(x5) cos(x1)+3.65887 sin(x2) sin(x5) cos(x1)+18.0372 sin(x3) sin(x5) cos(x1)+12.8227 sin(x4)   sin(x5)   cos(x1)+32.1636   sin(x5)   cos(x1)+14.4466   cos(x1)+0.77552 cos^3(x2)+0.753641   cos^3(x3)+1.29685   cos^3(x4)+1.04188   cos^3(x5)-1.50417 sin^3(x1)+1.48947 sin^3(x2)+0.62925 sin^3(x3)+0.958749 sin^3(x4)+0.220581 sin^3(x5)-1.90869   cos^2(x2)+5.06747   cos(x2)   cos^2(x3)+1.45896 cos^2(x3)-9.45157   cos(x2)   cos^2(x4)-1.20073   cos(x3)   cos^2(x4)+2.10493

cos^2(x4)-6.92812 cos(x2) cos^2(x5)+1.00938 cos(x3) cos^2(x5)+8.34976 cos(x4)
cos^2(x5)+4.12167 cos^2(x5)-4.78619 cos(x2) sin^2(x1)-3.81994 cos(x3)
sin^2(x1)-13.8024 cos(x4) sin^2(x1)-7.34562 cos(x5) sin^2(x1)-10.9339
sin^2(x1)-1.6675 cos(x2) sin^2(x2)-2.02331 cos(x3) sin^2(x2)+12.2598 cos(x4)
sin^2(x2)+14.2192 cos(x5) sin^2(x2)-1.93915 sin(x1) sin^2(x2)+4.06967
sin^2(x2)-5.75529 cos(x2) sin^2(x3)+3.2843 cos(x3) sin^2(x3)+3.51801 cos(x4)
sin^2(x3)+4.67355 cos(x5) sin^2(x3)-2.5468 sin(x1) sin^2(x3)+6.71373 sin(x2)
sin^2(x3)+0.702021 sin^2(x3)+8.76375 cos(x2) sin^2(x4)+5.2678 cos(x3)
sin^2(x4)-4.14309 cos(x4) sin^2(x4)-2.84299 cos(x5) sin^2(x4)+4.30677 sin(x1)
sin^2(x4)-12.2708 sin(x2) sin^2(x4)-2.31449 sin(x3) sin^2(x4)+0.0560541
sin^2(x4)+6.2403 cos(x2) sin^2(x5)+3.05769 cos(x3) sin^2(x5)-6.82122 cos(x4)
sin^2(x5)+2.58076 cos(x5) sin^2(x5)+0.927661 sin(x1) sin^2(x5)-12.1249 sin(x2)
sin^2(x5)-2.60022 sin(x3) sin^2(x5)+2.53606 sin(x4) sin^2(x5)-1.96069
sin^2(x5)-3.68782 cos(x2)+6.09038 cos^2(x2) cos(x3)+7.01176 cos(x2)
cos(x3)+1.06707 cos(x3)-10.7313 cos^2(x2) cos(x4)-1.98947 cos^2(x3)
cos(x4)-40.5569 cos(x2) cos(x4)-20.0592 cos(x2) cos(x3) cos(x4)-13.1832 cos(x3)
cos(x4)-1.47145 cos(x4)-8.26145 cos^2(x2) cos(x5)+1.28416 cos^2(x3)
cos(x5)+8.8007 cos^2(x4) cos(x5)-27.5937 cos(x2) cos(x5)-9.51423 cos(x2)
cos(x3) cos(x5)+2.10193 cos(x3) cos(x5)-35.1703 cos(x2) cos(x4) cos(x5)-0.420584
cos(x3) cos(x4) cos(x5)+17.7238 cos(x4) cos(x5)+2.95771 cos(x5)+2.92059
cos^2(x2) sin(x1)+3.52824 cos^2(x3) sin(x1)-3.32533 cos^2(x4)
sin(x1)+0.0537808 cos^2(x5) sin(x1)-6.79447 cos(x2) sin(x1)+20.6742 cos(x2)
cos(x3) sin(x1)+2.15625 cos(x3) sin(x1)-34.0021 cos(x2) cos(x4) sin(x1)-12.3631
cos(x3) cos(x4) sin(x1)-24.2409 cos(x4) sin(x1)-15.9919 cos(x2) cos(x5)
sin(x1)+4.30018 cos(x3) cos(x5) sin(x1)-6.43087 cos(x4) cos(x5) sin(x1)-6.34557
cos(x5) sin(x1)-2.01856 sin(x1)-4.1759 cos^2(x2) sin(x2)-3.48441 cos^2(x3)
sin(x2)+15.5001 cos^2(x4) sin(x2)+15.3542 cos^2(x5) sin(x2)-11.6972 sin^2(x1)
sin(x2)-23.5644 cos(x2) sin(x2)-13.9938 cos(x2) cos(x3) sin(x2)-17.8252 cos(x3)
sin(x2)-5.13474 cos(x2) cos(x4) sin(x2)+2.46122 cos(x3) cos(x4) sin(x2)+25.3956
cos(x4) sin(x2)+1.6668 cos(x2) cos(x5) sin(x2)+13.1848 cos(x3) cos(x5)
sin(x2)+69.3777 cos(x4) cos(x5) sin(x2)+36.0054 cos(x5) sin(x2)-23.6576 cos(x2)
sin(x1) sin(x2)-25.809 cos(x3) sin(x1) sin(x2)-0.773754 cos(x4) sin(x1)
sin(x2)+4.61271 cos(x5) sin(x1) sin(x2)-23.992 sin(x1) sin(x2)+0.229322
sin(x2)-4.2547 cos^2(x2) sin(x3)+4.56771 cos^2(x3) sin(x3)+6.54544 cos^2(x4)
sin(x3)+6.83117 cos^2(x5) sin(x3)-8.47039 sin^2(x1) sin(x3)+8.48565 sin^2(x2)
sin(x3)-22.2573 cos(x2) sin(x3)+7.66147 cos(x2) cos(x3) sin(x3)+8.32987 cos(x3)
sin(x3)-26.1849 cos(x2) cos(x4) sin(x3)+3.73758 cos(x3) cos(x4) sin(x3)+4.47567
cos(x4) sin(x3)-19.0877 cos(x2) cos(x5) sin(x3)+10.7058 cos(x3) cos(x5)
sin(x3)+29.8257 cos(x4) cos(x5) sin(x3)+12.8606 cos(x5) sin(x3)-19.1988 cos(x2)
sin(x1) sin(x3)+4.01637 cos(x3) sin(x1) sin(x3)-14.9445 cos(x4) sin(x1)

sin(x3)-7.49798 cos(x5) sin(x1) sin(x3)-12.9127 sin(x1) sin(x3)-11.9051 cos(x2) sin(x2) sin(x3)+0.0270216 cos(x3) sin(x2) sin(x3)+36.0087 cos(x4) sin(x2) sin(x3)+38.9931 cos(x5) sin(x2) sin(x3)-4.0791 sin(x1) sin(x2) sin(x3)+13.6657 sin(x2) sin(x3)+1.23096 sin(x3)+13.2666 cos^2(x2) sin(x4)+6.1685 cos^2(x3) sin(x4)-0.611044   cos^2(x4)   sin(x4)-0.518361   cos^2(x5)   sin(x4)-0.156292 sin^2(x1) sin(x4)-11.2489 sin^2(x2) sin(x4)-4.1508 sin^2(x3) sin(x4)+20.2192 cos(x2) sin(x4)+40.6212 cos(x2) cos(x3) sin(x4)+11.578 cos(x3) sin(x4)-0.379336 cos(x2)   cos(x4)   sin(x4)-2.10699   cos(x3)   cos(x4)   sin(x4)-19.2483   cos(x4) sin(x4)+8.64006 cos(x2) cos(x5) sin(x4)+8.61802 cos(x3) cos(x5) sin(x4)+0.314232 cos(x4)   cos(x5)   sin(x4)-9.59129   cos(x5)   sin(x4)+32.2971   cos(x2)   sin(x1) sin(x4)+17.0056 cos(x3) sin(x1) sin(x4)-12.3467 cos(x4) sin(x1) sin(x4)-0.914383 cos(x5)   sin(x1)   sin(x4)+2.17029   sin(x1)   sin(x4)-31.5856   cos(x2)   sin(x2) sin(x4)-35.4876 cos(x3) sin(x2) sin(x4)-32.2002 cos(x4) sin(x2) sin(x4)-29.848 cos(x5)   sin(x2)   sin(x4)-42.3074   sin(x1)   sin(x2)   sin(x4)-52.2806   sin(x2) sin(x4)-0.824576 cos(x2) sin(x3) sin(x4)+6.88731 cos(x3) sin(x3) sin(x4)-16.763 cos(x4) sin(x3) sin(x4)-14.2123 cos(x5) sin(x3) sin(x4)-5.38494 sin(x1) sin(x3) sin(x4)-33.5755   sin(x2)   sin(x3)   sin(x4)-15.9248   sin(x3)   sin(x4)-0.982303 sin(x4)+10.9011 cos^2(x2) sin(x5)+6.15169 cos^2(x3) sin(x5)-1.84462 cos^2(x4) sin(x5)+7.75697 cos^2(x5) sin(x5)-4.05151 sin^2(x1) sin(x5)-7.1589 sin^2(x2) sin(x5)-2.40949 sin^2(x3) sin(x5)+5.58682 sin^2(x4) sin(x5)+16.1832 cos(x2) sin(x5)+36.6137 cos(x2) cos(x3) sin(x5)+13.772 cos(x3) sin(x5)-6.30811 cos(x2) cos(x4) sin(x5)-2.04133 cos(x3) cos(x4) sin(x5)-20.3969 cos(x4) sin(x5)+12.3944 cos(x2) cos(x5) sin(x5)+22.449 cos(x3) cos(x5) sin(x5)+11.5555 cos(x4) cos(x5) sin(x5)+15.2842 cos(x5) sin(x5)+19.6399 cos(x2) sin(x1) sin(x5)+13.0393 cos(x3) sin(x1)   sin(x5)-16.0166   cos(x4)   sin(x1)   sin(x5)+10.4898   cos(x5)   sin(x1) sin(x5)+0.595998 sin(x1) sin(x5)-25.8129 cos(x2) sin(x2) sin(x5)-28.0372 cos(x3) sin(x2)   sin(x5)-27.9899   cos(x4)   sin(x2)   sin(x5)+4.10122   cos(x5)   sin(x2) sin(x5)-29.8687 sin(x1) sin(x2) sin(x5)-36.8172 sin(x2) sin(x5)-2.30976 cos(x2) sin(x3)   sin(x5)+10.9269   cos(x3)   sin(x3)   sin(x5)-17.5495   cos(x4)   sin(x3) sin(x5)+6.22595 cos(x5) sin(x3) sin(x5)-3.06398 sin(x1) sin(x3) sin(x5)-18.7023 sin(x2)   sin(x3)   sin(x5)-7.91324   sin(x3)   sin(x5)+30.7279   cos(x2)   sin(x4) sin(x5)+18.9796 cos(x3) sin(x4) sin(x5)-18.3111 cos(x4) sin(x4) sin(x5)+6.97693 cos(x5) sin(x4) sin(x5)+14.0771 sin(x1) sin(x4) sin(x5)-38.2439 sin(x2) sin(x4) sin(x5)-3.07525   sin(x3)   sin(x4)   sin(x5)+5.74291   sin(x4)   sin(x5)+0.742202 sin(x5)+0.573443)/(2.72438   cos^3(x1)+11.4454   cos(x2)   cos^2(x1)+10.7032 cos(x3) cos^2(x1)+8.38975 cos(x4) cos^2(x1)+10.7731 cos(x5) cos^2(x1)-4.2981 sin(x1) cos^2(x1)+8.83798 sin(x2) cos^2(x1)+7.88459 sin(x3) cos^2(x1)+11.3526 sin(x4)   cos^2(x1)+7.10081   sin(x5)   cos^2(x1)+12.8146   cos^2(x1)+3.45595 cos^2(x2) cos(x1)+5.90621 cos^2(x3) cos(x1)+3.69777 cos^2(x4) cos(x1)+8.21337 cos^2(x5) cos(x1)-7.4797 sin^2(x1) cos(x1)+7.58376 sin^2(x2) cos(x1)+5.1335 sin^2(x3) cos(x1)+7.34194 sin^2(x4) cos(x1)+2.82634 sin^2(x5) cos(x1)+17.5365 cos(x2) cos(x1)+9.16889 cos(x2) cos(x3) cos(x1)+18.3146 cos(x3) cos(x1)+22.7034 cos(x2) cos(x4) cos(x1)+21.6063 cos(x3) cos(x4) cos(x1)+17.9683 cos(x4) cos(x1)+33.677 cos(x2) cos(x5) cos(x1)+26.8653 cos(x3) cos(x5) cos(x1)+13.1353 cos(x4) cos(x5) cos(x1)+25.5216 cos(x5) cos(x1)-32.9277 cos(x2) sin(x1) cos(x1)-28.3916 cos(x3) sin(x1) cos(x1)-25.051 cos(x4) sin(x1) cos(x1)-16.9164 cos(x5) sin(x1) cos(x1)-33.0357 sin(x1) cos(x1)+24.9099 cos(x2) sin(x2) cos(x1)+24.4649 cos(x3) sin(x2) cos(x1)+5.72191 cos(x4) sin(x2) cos(x1)+5.19343 cos(x5) sin(x2) cos(x1)-1.46903 sin(x1) sin(x2) cos(x1)+29.5448 sin(x2) cos(x1)+11.6969 cos(x2) sin(x3) cos(x1)+6.08861 cos(x3) sin(x3) cos(x1)+2.2328 cos(x4) sin(x3) cos(x1)+7.97292 cos(x5) sin(x3) cos(x1)-18.6141 sin(x1) sin(x3) cos(x1)+18.2424 sin(x2) sin(x3) cos(x1)+15.1928 sin(x3) cos(x1)-2.06534 cos(x2) sin(x4) cos(x1)+3.41136 cos(x3) sin(x4) cos(x1)+16.3108 cos(x4) sin(x4) cos(x1)+21.9159 cos(x5) sin(x4) cos(x1)-16.81 sin(x1) sin(x4) cos(x1)+47.5623 sin(x2) sin(x4) cos(x1)+26.5109 sin(x3) sin(x4) cos(x1)+22.3393 sin(x4) cos(x1)-16.834 cos(x2) sin(x5) cos(x1)-8.16465 cos(x3) sin(x5) cos(x1)+6.64268 cos(x4) sin(x5) cos(x1)+1.63099 cos(x5) sin(x5) cos(x1)-32.6418 sin(x1) sin(x5) cos(x1)+37.0864 sin(x2) sin(x5) cos(x1)+12.8528 sin(x3) sin(x5) cos(x1)+8.08371 sin(x4) sin(x5) cos(x1)+4.42844 sin(x5) cos(x1)+8.03971 cos(x1)-0.0110681 cos^3(x2)+1.21018 cos^3(x3)+1.81246 cos^3(x4)+1.9348 cos^3(x5)+1.18414 sin^3(x1)+1.99702 sin^3(x2)+1.73807 sin^3(x3)+0.644592 sin^3(x4)+1.08223 sin^3(x5)-2.6183 cos^2(x2)-1.46752 cos(x2) cos^2(x3)+3.89035 cos^2(x3) +14.0561 cos(x2) cos^2(x4)+11.3817 cos(x3) cos^2(x4)+11.4477 cos^2(x4)+13.1062 cos(x2) cos^2(x5)+9.90418 cos(x3) cos^2(x5)+9.50354 cos(x4) cos^2(x5)+10.0264 cos^2(x5)-5.88812 cos(x2) sin^2(x1)-3.11041 cos(x3) sin^2(x1)+6.62148 cos(x4) sin^2(x1)+1.35798 cos(x5) sin^2(x1)-0.723922 sin^2(x1)+11.6569 cos(x2) sin^2(x2)+13.2482 cos(x3) sin^2(x2)+6.27952 cos(x4) sin^2(x2)+4.3254 cos(x5) sin^2(x2)+10.9996 sin(x1) sin^2(x2)+14.7089 sin^2(x2)+7.02484 cos(x2) sin^2(x3)+2.70123 cos(x3) sin^2(x3)+5.84674 cos(x4) sin^2(x3)+5.01198 cos(x5) sin^2(x3)+4.17715 sin(x1) sin^2(x3)+7.93308 sin(x2) sin^2(x3)+8.20028 sin^2(x3)-8.49876 cos(x2) sin^2(x4)-3.78891 cos(x3) sin^2(x4)+4.69911 cos(x4) sin^2(x4)+2.78736 cos(x5) sin^2(x4)-5.93008 sin(x1) sin^2(x4)+15.5012 sin(x2) sin^2(x4)+5.07385 sin(x3) sin^2(x4)+0.642924 sin^2(x4)-7.54891 cos(x2) sin^2(x5)-2.31137 cos(x3) sin^2(x5)+5.50769 cos(x4) sin^2(x5)+0.717851 cos(x5) sin^2(x5)-2.94363 sin(x1) sin^2(x5)+16.2037 sin(x2) sin^2(x5)+5.37491 sin(x3) sin^2(x5)+0.849318 sin(x4) sin^2(x5)+2.06419 sin^2(x5)+2.55732 cos(x2)-5.65537 cos^2(x2) cos(x3)-12.4689 cos(x2) cos(x3)+4.59281 cos(x3)+8.73171 cos^2(x2) cos(x4)+9.16449 cos^2(x3) cos(x4)+35.9017 cos(x2) cos(x4)+22.035 cos(x2) cos(x3) cos(x4)+32.8049 cos(x3) cos(x4)+12.0112 cos(x4)+7.80568 cos^2(x2) cos(x5)+7.1191 cos^2(x3) cos(x5)+9.34372 cos^2(x4) cos(x5)+29.491 cos(x2) cos(x5)+17.0145 cos(x2) cos(x3) cos(x5)+23.0592 cos(x3) cos(x5)+49.284 cos(x2) cos(x4) cos(x5)+35.4507 cos(x3) cos(x4) cos(x5)+32.9828 cos(x4) cos(x5)+9.13108 cos(x5)-10.6404 cos^2(x2) sin(x1)-3.81798 cos^2(x3) sin(x1)+6.28925 cos^2(x4) sin(x1)+3.3028 cos^2(x5) sin(x1)-35.4295 cos(x2) sin(x1)-43.6335 cos(x2) cos(x3) sin(x1)-24.353 cos(x3) sin(x1)+4.75714 cos(x2) cos(x4) sin(x1)+2.59261 cos(x3) cos(x4) sin(x1)+9.70891 cos(x4) sin(x1)-7.58897 cos(x2) cos(x5) sin(x1)-9.72509 cos(x3) cos(x5) sin(x1)+12.1671 cos(x4) cos(x5) sin(x1)-4.1553 cos(x5) sin(x1)-2.64083 sin(x1)+6.59552 cos^2(x2) sin(x2)+10.6357 cos^2(x3) sin(x2)+3.06749 cos^2(x4) sin(x2)+2.36505 cos^2(x5) sin(x2)+9.73076 sin^2(x1) sin(x2)+37.2491 cos(x2) sin(x2)+23.1853 cos(x2) cos(x3) sin(x2)+43.7976 cos(x3) sin(x2)+27.2826 cos(x2) cos(x4) sin(x2)+30.4449 cos(x3) cos(x4) sin(x2)+29.6774 cos(x4) sin(x2)+23.0701 cos(x2) cos(x5) sin(x2)+21.9162 cos(x3) cos(x5) sin(x2)-2.91076 cos(x4) cos(x5) sin(x2)+20.909 cos(x5) sin(x2)+14.6564 cos(x2) sin(x1) sin(x2)+24.1447 cos(x3) sin(x1) sin(x2)+18.1359 cos(x4) sin(x1) sin(x2)+12.3976 cos(x5) sin(x1) sin(x2)+30.1048 sin(x1) sin(x2)+15.5687 sin(x2)+1.21123 cos^2(x2) sin(x3)+0.591966 cos^2(x3) sin(x3)+5.16073 cos^2(x4) sin(x3)+4.85967 cos^2(x5) sin(x3)+2.34999 sin^2(x1) sin(x3)+9.02335 sin^2(x2) sin(x3)+10.8295 cos(x2) sin(x3)-10.6107 cos(x2) cos(x3) sin(x3)+2.37444 cos(x3) sin(x3)+24.0474 cos(x2) cos(x4) sin(x3)+12.5682 cos(x3) cos(x4) sin(x3)+22.1001 cos(x4) sin(x3)+20.3607 cos(x2) cos(x5) sin(x3)+8.00039 cos(x3) cos(x5) sin(x3)+10.1895 cos(x4) cos(x5) sin(x3)+16.4316 cos(x5) sin(x3)-10.2693 cos(x2) sin(x1) sin(x3)-15.8547 cos(x3) sin(x1) sin(x3)+9.44911 cos(x4) sin(x1) sin(x3)+2.9169 cos(x5) sin(x1) sin(x3)-3.0418 sin(x1) sin(x3)+27.2766 cos(x2) sin(x2) sin(x3)+24.3514 cos(x3) sin(x2) sin(x3)+11.8108 cos(x4) sin(x2) sin(x3)+8.84581 cos(x5) sin(x2) sin(x3)+18.6331 sin(x1) sin(x2) sin(x3)+34.5607 sin(x2) sin(x3)+7.23458 sin(x3)-14.1998 cos^2(x2) sin(x4)-3.46755 cos^2(x3) sin(x4)+5.46367 cos^2(x4) sin(x4)+4.41568 cos^2(x5) sin(x4)-6.0876 sin^2(x1) sin(x4)+19.4648 sin^2(x2) sin(x4)+8.73255 sin^2(x3) sin(x4)-36.6958 cos(x2) sin(x4)-45.8932 cos(x2) cos(x3) sin(x4)-16.9393 cos(x3) sin(x4)-8.64767 cos(x2) cos(x4) sin(x4)+2.48679 cos(x3) cos(x4) sin(x4)+16.9445 cos(x4) sin(x4)-16.1449 cos(x2) cos(x5) sin(x4)-6.69902 cos(x3) cos(x5) sin(x4)+10.4605 cos(x4) cos(x5) sin(x4)+7.74261 cos(x5) sin(x4)-56.7609 cos(x2) sin(x1) sin(x4)-35.1356 cos(x3) sin(x1) sin(x4)-5.27904 cos(x4) sin(x1) sin(x4)-17.0229 cos(x5) sin(x1) sin(x4)-29.8731 sin(x1) sin(x4)+29.0941 cos(x2) sin(x2) sin(x4)+37.9171 cos(x3) sin(x2) sin(x4)+43.9686 cos(x4) sin(x2) sin(x4)+40.3782 cos(x5) sin(x2) sin(x4)+33.1862 sin(x1) sin(x2) sin(x4)+62.4105 sin(x2) sin(x4)-10.0489 cos(x2) sin(x3) sin(x4)-6.87839 cos(x3) sin(x3) sin(x4)+18.9059 cos(x4) sin(x3) sin(x4)+15.2478 cos(x5) sin(x3) sin(x4)-9.1548 sin(x1) sin(x3) sin(x4)+49.8787 sin(x2) sin(x3) sin(x4)+16.7209 sin(x3) sin(x4)+2.265 sin(x4)-13.0593 cos^2(x2)

sin(x5)-3.07549 cos^2(x3) sin(x5)+5.85253 cos^2(x4) sin(x5)+1.05987 cos^2(x5)
sin(x5)-2.30089 sin^2(x1) sin(x5)+17.8593 sin^2(x2) sin(x5)+7.87541 sin^2(x3)
sin(x5)-1.05261 sin^2(x4) sin(x5)-34.412 cos(x2) sin(x5)-43.696 cos(x2) cos(x3)
sin(x5)-16.3452 cos(x3) sin(x5)-7.65356 cos(x2) cos(x4) sin(x5)+1.93598 cos(x3)
cos(x4) sin(x5)+17.2477 cos(x4) sin(x5)-14.7328 cos(x2) cos(x5) sin(x5)-10.5094
cos(x3) cos(x5) sin(x5)+6.47734 cos(x4) cos(x5) sin(x5)-2.00174 cos(x5)
sin(x5)-47.2278 cos(x2) sin(x1) sin(x5)-29.9674 cos(x3) sin(x1) sin(x5)-2.88797
cos(x4) sin(x1) sin(x5)-18.3664 cos(x5) sin(x1) sin(x5)-25.2985 sin(x1)
sin(x5)+26.326 cos(x2) sin(x2) sin(x5)+36.2671 cos(x3) sin(x2) sin(x5)+43.6433
cos(x4) sin(x2) sin(x5)+22.995 cos(x5) sin(x2) sin(x5)+28.5044 sin(x1) sin(x2)
sin(x5)+56.1764 sin(x2) sin(x5)-9.10425 cos(x2) sin(x3) sin(x5)-8.24901 cos(x3)
sin(x3) sin(x5)+19.1853 cos(x4) sin(x3) sin(x5)+5.85661 cos(x5) sin(x3)
sin(x5)-8.25217 sin(x1) sin(x3) sin(x5)+42.2985 sin(x2) sin(x3) sin(x5)+13.1587
sin(x3) sin(x5)-39.072 cos(x2) sin(x4) sin(x5)-20.7017 cos(x3) sin(x4)
sin(x5)+10.7107 cos(x4) sin(x4) sin(x5)-5.40577 cos(x5) sin(x4) sin(x5)-26.8966
sin(x1) sin(x4) sin(x5)+47.7992 sin(x2) sin(x4) sin(x5)+8.66295 sin(x3) sin(x4)
sin(x5)-7.22002 sin(x4) sin(x5)+1.79992 sin(x5)+1.67674)

References

[1] Guedes Soares, C. 1990. Stochastic models of load effects for the primary ship structure. Struct. Saf. 8(1-4): 353–368.
[2] Leheta, H.W. and Mansour, A.E. 2002. Reliability-based method for optimal structural design of stiffened panels. Mar. Struct. 10(5): 323–352.
[3] Amlashi, H.K.K. and Moan, T. 2009. Ultimate strength analysis of a bulk carrier hull girder under alternate hold loading condition, Part 2: Stress distribution in the double bottom and simplified approaches. Mar. Struct. 22(3): 522–544.
[4] Sekulski, Z. 2011. Multi-objective optimization of high speed vehicle-passenger catamaran by genetic algorithm: Part II Computational simulations. Polish Marit. Res. 18(3): 3–30.
[5] Fu, S.Y., Huang, H.Y. and Lin, Z.X. 2012. Collaborative optimization of container ship on static and dynamic responses. Procedia Eng. 31: 613–621.
[6] Bayatfar, A., Amrane, A. and Rigo, P. 2013. Towards a ship structural optimisation methodology at early design stage. Int. J. Eng. Res. Dev. 9(6): 76–90.
[7] Shi, X., Teixeira, A.P., Zhang, J. and Soares, C.G. 2014. Structural reliability analysis based on probabilistic response modelling using the Maximum Entropy Method. Eng. Struct. 70: 106–116.
[8] Campanile, A., Piscopo, V. and Scamardella, A. 2014. Statistical properties of bulk carrier longitudinal strength. Mar. Struct. 39: 438–462.
[9] Sekulski, Z. 2014. Ship hull structural multiobjective optimization by evolutionary algorithm. J. Sh. Res. 58(2): 45–69.
[10] Um, T.S. and Roh, M.I. 2015. Optimal dimension design of a hatch cover for lightening a bulk carrier. Int. J. Nav. Archit. Ocean Eng. 7(2): 270–287.
[11] Anyfantis, K.N. 2019. Evaluating the influence of geometric distortions to the buckling capacity of stiffened panels. Thin-Walled Struct. 140: 450–465.

[12] Song, Z.J., Xu, M.C., Moan, T. and Pan, J. 2018. Dimensional and similitude analysis of stiffened panels under longitudinal compression considering buckling behaviours. Ocean Eng. 87: 106–188.

[13] Chen, S., Dong, T. and Shui, X. 2019. Simultaneous distribution and sizing optimization for stiffeners with an improved genetic algorithm with two-level approximation. Engineering Optimization 51(11): 1845–1866.

[14] Raikunen, J., Avi, E., Remes, H., Romanoff, J., Lillemäe-Avi, I. and Niemelä, A. 2019. Optimisation of passenger ship structures in concept design stage. Ships Offshore Struct. 1–14.

[15] Lee, J.C., Shin, S.C. and Kim, S.Y. 2015. An optimal design of wind turbine and ship structure based on neuro-response surface method. Int. J. Nav. Archit. Ocean Eng. 7(4): 750–769.

[16] Kim, S.J. 2012. A Benchmark Study on Ultimate Strength Analysis Methods of Ship Stiffened Panel. M.S. Thesis, Pusan National University (in Korean).

CHAPTER 12

Optimization of Weld Bead Geometry for TIG Welding Process

Fatih Turhan,[1] *Cansu Yildirim*[2] *and Levent Aydin*[2,*]

Introduction

Welding is an essential process in the manufacturing of components, assemblies, or complete machines across the engineering spectrum. Tungsten inert gas (TIG) welding is a well-known and the most frequently used method of welding processes in ships, bridges, and welding of stainless steels. TIG weld quality is strongly characterized by the weld pool geometry which has several quality responses such as Front Height (FH), Front Width (FW), Back Height (BH), Back Width (BW) and area of penetration. Weld pool geometry plays a vital role in determining the mechanical properties of the weld. Therefore, it is crucial to select the welding process parameters for obtaining an optimal weld pool geometry [1].

The organization of this chapter is as follows:

This study consists of six sub-sections. In the first section, the importance of the TIG weld and cross-sectional geometry of weld bead is discussed with the relevant literature; moreover, the purpose of this study is briefly explained. Secondly, the TIG weld, which is one of the bonding methods, is

[1] Superrad Metal San. ve Tic. Ltd. Şti Istanbul, Turkey.
  Email: fatih_turhan@hotmail.com
[2] İzmir Katip Çelebi University, Department of Mechanical Eng. İzmir, Turkey.
  Email: cansuyu94@gmail.com
* Corresponding author: leventaydinn@gmail.com

described along with the equipment used. A more detailed description of Regression Analysis is given in the third section. After that, the functions prepared for our engineering problem and the results of the regression analyzes made with these functions are given. In the fifth section, general information about optimization is given, and the "Differential Evolution" method is described in detail. In the final section of the chapter, the data obtained as a result of optimization studies are discussed. In addition to this, the effects of the input parameters of welding on the objective functions are examined.

Literature Survey

The section geometry of the weld seam leads among the parameters affecting TIG weld quality, and it plays a vital role in helping to determine the mechanical properties of the weld. Nanda et al. [2] investigated parameters such as top width, top height, and penetration which make up the section geometry of the welding seam. It is stated that they are essential factors for welding quality. Nonlinear equations can model the TIG process as in the majority of arc weld operations. Besides, TIG welding is characterized by low heat input, low spatter, and weld cleaning. In addition, a tungsten electrode, which is not used to the protective application, and inert argon gas have been used in the TIG welding operation. When the literature is overviewed, it is seen that the researchers focused on various aspects of modeling, simulation, and process optimization in TIG welding. The mathematical expression of the weld geometry has been a significant concern for researchers because it is a result of the entire thermal cycles of a weld.

Regarding this fact, several researchers have used some traditional regression analysis approaches and have made several attempts to model the welding process. Narang et al. [3], modeled the weld pool profile properties in the macrostructure region of weld with "Fuzzy Logic Simulation." In the modeling process, the geometry of the weld seam section geometry along with the shape of the heat-affected zone was calculated by taking into account the TIG welding process parameters such as arc speed, weld current, and arc length.

In previous studies, both linear and nonlinear mathematical models have been performed based on the collected experimental data in a particular mode (e.g., full factorial or fractional factorial designs of experiments). Tarng et al. [4] used "Neural Networks" in predicting front width, back width, front height and rear height of TIG welding, and also both "back-propagation" and "counter-propagation" to create complex relationships between welding process parameters and weld seam properties. Experimental results show that "counter-propagation" network for TIG welding has a better learning ability than "back-propagation" network. In contrast, Yang et al. [5] have used curvilinear regression equations to model the properties of welded stitch cross-sectional geometry, while the back-propagation network has a better generalization capability than the counter-propagation network. However, only the correlation coefficients are reported in this study. This study was conducted to investigate the relationship between the standard deviation of the differences between estimated and measured values by the correlation coefficient and the dissolution rate, total fusion area, penetration, weld seam height, and weld pool.

The selection of optimum welding process parameters is crucial since they affect the quality, cost, and strength of welding operation. Therefore, the researchers have focused on optimal solutions for the engineering problems in this area. For example, Lee et al. [6] have found the objective functions of welding parameters by multiple regression analysis in a gas metal arc welding process. The regression model is obtained by correlating the back-height parameter of the weld seam with the welding process parameters applied by the inverse transform. As a result, the maximum prediction error rate in the forward process is less than 9.5%, and the prediction error rate in the reverse process of the source process parameters is less than 6.5%. Kim et al. [7] developed linear and non-linear multiple regression models to determine the relationship between process variables and weld seam penetration for robotic CO_2 arc welding. The developed mathematical model predicted penetration with reasonable accuracy. Dutta et al. [8] modeled a TIG welding process using a classical linear regression technique, Advance Back-Propagation Neural Network (ABPNN) and Genetic Algorithms-Neural Network (GA-ANN). After obtaining the appropriate models, their performances have been compared with some test cases. It has been found that "NN" based approaches are more applicable than conventional regression analysis for the modeling of the experimental process. Joby et al. [9] studied the tensile strength of specimens to determine the weld seam quality. They prepared regression equations for these processes using Response Surface Method (RSM) and then optimized the design variables by using Genetic Algorithm (GA) and Simulated Annealing (SA) algorithms to select the ideal source parameters. These models gave satisfactory results as a part of the optimization procedure. In the validation test, it was seen that GA and SA could be used effectively when the input process parameters were specified. Dey et al. [10] performed the welding process on an Al-1100 aluminum plate using an electron beam welding machine. Then, they performed a regression analysis to determine the input-output relationships of the process. In their study, it was observed that when the maximum penetration condition was fulfilled, the welding field decreased to a minimum. Moreover, the constrained optimization problem was solved by using the Genetic Algorithm with a penalty function approach, and the optimal section geometry of the weld seam could be determined.

Objectives and Motivation of the Study

The arc weld is an extremely complex process that occurs at high temperatures and causes permanent internal stress in the target material. TIG weld is one of the most critical production operations in a wide range of application areas requiring high precision such as railway, vessels, bridges, construction, automotive, aircraft, and nuclear reactors. The weld seam geometry plays a vital role in determining the mechanical properties of the weld. For this reason, it is essential to select the welding process parameters in order to obtain the optimum weld segment geometry. The primary purpose of many engineering applications is to reduce cost and weight without sacrificing mechanical properties. It should also be noted that optimization has a crucial role in the design of all complex systems, as well as in engineering applications.

The objectives to be achieved in this study are:

- Development of mathematical models of TIG weld input parameters and associated welded suture geometries.
- Determine optimum weld input parameters to achieve maximum penetration and minimum weld seam cross-section area.
- Investigating the effects of TIG welding input parameters on the weld seam.

In this study, the regression model was created by utilizing the welding process data [11], weld seam section geometry examined the effects of welding input parameters [12], the welding stitch section geometry was used to construct and optimize the function [13], the performance factor was accounted for depending on the welding input parameters [14].

Definition of the Problem

It is assumed that the quality of the prescribed TIG weld operation is strongly related to the weld seam section geometry with various quality features, as shown in Figure 12.1. These are front width (FW), front height (FH), •back width (BW), back height (BH), penetration depth (BH + t) [1].

In this study, the area of the section geometry of the weld seam under the maximum penetration conditions has been calculated. For this aim, then five different optimization problems have been carried out and presented in Table 12.1.

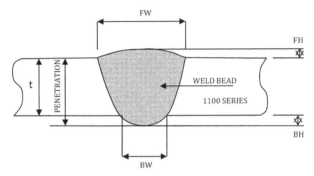

Figure 12.1 Scheme of the cross-section geometry of weld seam [1].

Table 12.1 Objective functions, constraints and design variables of the optimization problems [1].

Problem	Objective	Constraints	Design Variables
1	Front Width (FW) Minimization	• FH < 1+0.1 FW (ISO 5817:2014) • Welding Speed (a) Є {24, 35, 46}; • Wire Speed (b) Є {1.5, 2, 2.5};	
2	Front Height (FH) Minimization	• Gas Flow Rate (c) Є {30, 70}; • Gap Distance (d) Є {2.4, 3.2}; • Welding Current (e) Є {80, 95, 110}	• Welding Speed (a)
3	Back Width (BW) Minimization	• BH < 1+0.1 BW (ISO 5817:2014) • Welding Speed (a) Є {24, 35, 46}; • Wire Speed (b) C {1.5, 2, 2.5};	• Wire Speed (b) • Gas Flow Rate (c)
4	Back Height (BH) Maximization	• Gas Flow Rate (c) Є {30, 70}; • Gap Distance (d) C {2.4, 3.2}; • Welding Current (e) Є {80, 95, 110}	• Gap Distance (d)
5	Weld Bead Cross-Sectional Area Minimization	• BH > (0.1–0.7) • Welding Speed (a) Є {24, 35, 46}; • Wire Speed (b) Є {1.5, 2, 2.5}; • Gas Flow Rate (c) Є {30, 70}; • Gap Distance (d) Є {2.4, 3.2}; • Welding Current (e) Є {80, 95, 110}	• Welding Current (e)

Regression Analysis

In this study, the experimental data for the welding process obtained by Tarng [3] was used (Table 12.2 and Table 12.3). In addition, a mathematical model of the process has been developed to determine the optimal values of the TIG welding parameters. For this purpose, multiple nonlinear regression models are used. Then, the obtained results are used to optimize the process and determine the effect of the parameters on the

Table 12.2 Training experimental set for TIG welding operation [2].

No.	INPUTS					OUTPUTS			
	Welding Speed (cm/min)	Wire Speed (cm/min)	Cleaning (%)	Gap (mm)	Current (A)	Front Height (mm)	Front Width (mm)	Back Height (mm)	Back Width (mm)
1	24	1.5	30	2.4	80	−0.149	6.09	0.672	5.664
2	24	1.5	30	3.2	80	0.027	6.411	0.412	5.197
3	24	1.5	70	2.4	80	−0.179	7.432	0.593	7.058
4	24	1.5	70	3.2	80	−0.306	7.287	0.63	6.895
5	24	2.5	30	2.4	80	0.155	6.676	0.743	5.96
6	24	2.5	30	3.2	80	0.099	6.824	0.803	5.732
7	24	2.5	70	2.4	80	−0.129	7.009	0.878	6.989
8	24	2.5	70	3.2	80	−0.077	7.46	0.82	7.809
9	24	1.5	30	2.4	95	−0.017	8.664	0.437	8.75
10	24	1.5	30	3.2	95	−0.25	8.782	0.593	9.993
11	24	1.5	70	2.4	95	−0.553	9.757	0.852	9.993
12	24	1.5	70	3.2	95	−0.42	10.374	0.736	10.687
13	24	2.5	30	2.4	95	−0.345	9.783	0.965	10.237
14	24	2.5	30	3.2	95	−0.043	8.803	0.654	9.076
15	24	2.5	70	2.4	95	−0.134	9.75	0.798	9.465
16	24	2.5	70	3.2	95	−0.168	10.348	0.708	10.193
17	24	1.5	30	2.4	110	−0.599	11.348	0.805	11.679
18	24	1.5	30	3.2	110	−0.745	11.491	1.1	11.848
19	24	1.5	70	2.4	110	−0.254	11.237	0.47	12
20	24	1.5	70	3.2	110	−0.683	12.946	0.945	13.921
21	24	2.5	30	2.4	110	−0.232	9.338	0.866	10.611
22	24	2.5	30	3.2	110	−0.557	12.348	1.139	12.403
23	24	2.5	70	2.4	110	−0.623	11.767	1.128	12.86
24	24	2.5	70	3.2	110	−0.617	12.533	1.084	13.346
25	35	1.5	30	2.4	80	0.123	5.355	0.245	4.104
26	35	1.5	30	3.2	80	0.108	5.173	0.34	3.418
27	35	1.5	70	2.4	80	−0.044	5.833	0.51	4.875

Table 12.2 contd. ...

... Table 12.2 contd.

No.	INPUTS					OUTPUTS			
	Welding Speed (cm/min)	Wire Speed (cm/min)	Cleaning (%)	Gap (mm)	Current (A)	Front Height (mm)	Front Width (mm)	Back Height (mm)	Back Width (mm)
28	35	1.5	70	3.2	80	−0.09	5.831	0.502	5.082
29	35	2.5	30	2.4	80	0.251	5.656	0.557	4.37
30	35	2.5	30	3.2	80	0.23	5.562	0.593	3.948
31	35	2.5	70	2.4	80	0.18	5.711	0.45	5.085
32	35	2.5	70	3.2	80	0.12	5.85	0.626	4.989
33	35	1.5	30	2.4	95	−0.213	6.348	0.458	5.874
34	35	1.5	30	3.2	95	−0.19	6.992	0.447	6.74
35	35	1.5	70	2.4	95	−0.152	7.163	0.464	6.994
36	35	1.5	70	3.2	95	−0.213	7.25	0.504	7.019
37	35	2.5	30	2.4	95	−0.164	7.288	0.715	6.724
38	35	2.5	30	3.2	95	−0.113	6.966	0.746	6.433
39	35	2.5	70	2.4	95	−0.107	7.055	0.696	7.24
40	35	2.5	70	3.2	95	−0.018	7.549	0.591	7.166
41	35	1.5	30	2.4	110	−0.575	8.337	0.766	8.763
42	35	1.5	30	3.2	110	−0.267	8.605	0.506	8.58
43	35	1.5	70	2.4	110	−0.385	9.109	0.672	9.652
44	35	1.5	70	3.2	110	−0.564	9.67	0.743	9.952
45	35	2.5	30	2.4	110	−0.556	8.756	1.011	8.853
46	35	2.5	30	3.2	110	−0.188	9.442	0.666	9.614
47	35	2.5	70	2.4	110	−0.309	9.015	0.784	9.041
48	35	2.5	70	3.2	110	−0.318	9.297	0.785	9.47
49	46	1.5	30	2.4	80	0.357	4.982	0.001	2.255
50	46	1.5	30	3.2	80	0.168	4.898	0.277	2.998
51	46	1.5	70	2.4	80	0.088	5.02	0.281	3.302
52	46	1.5	70	3.2	80	0.09	4.423	0.42	3.172
53	46	2.5	30	2.4	80	0.39	4.78	0.062	1.33
54	46	2.5	30	3.2	80	0.487	4.992	0.139	1.6

Table 12.2 contd. ...

... Table 12.2 contd.

No.	INPUTS					OUTPUTS			
	Welding Speed (cm/min)	Wire Speed (cm/min)	Cleaning (%)	Gap (mm)	Current (A)	Front Height (mm)	Front Width (mm)	Back Height (mm)	Back Width (mm)
55	46	2.5	70	2.4	80	0.38	5.231	0.397	2.817
56	46	2.5	70	3.2	80	0.394	5.337	0.378	3.041
57	46	1.5	30	2.4	95	−0.321	5.847	0.44	5.332
58	46	1.5	30	3.2	95	−0.152	5.704	0.386	5.35
59	46	1.5	70	2.4	95	−0.155	5.967	0.445	5.415
60	46	1.5	70	3.2	95	−0.09	5.892	0.399	5.319
61	46	2.5	30	2.4	95	−0.236	5.984	0.696	5.531
62	46	2.5	30	3.2	95	0.067	6.03	0.575	5.636
63	46	2.5	70	2.4	95	−0.075	5.562	0.816	4.835
64	46	2.5	70	3.2	95	0.138	6.546	0.575	6.285
65	46	1.5	30	2.4	110	−0.217	6.092	0.359	6.419
66	46	1.5	30	3.2	110	−0.339	7.335	0.619	7.52
67	46	1.5	70	2.4	110	−0.249	7.719	0.492	7.706
68	46	1.5	70	3.2	110	−0.396	7.633	0.458	7.601
69	46	2.5	30	2.4	110	−0.01	6.396	0.536	6.197
70	46	2.5	30	3.2	110	0.074	6.863	0.484	6.072
71	46	2.5	70	2.4	110	−0.201	7.052	0.658	7.48
72	46	2.5	70	3.2	110	−0.385	7.759	0.798	7.917

objective functions. In this study, when the cross-section geometry of the welding seam is modeled, the results of standard nonlinear multiple regression analysis and "Artificial Neural Network" (ANN) method are used together. For this reason, the data of 108 experimental works carried out by Tarng et al. [4] were divided into two parts as training (72) and testing (36) data. Regression modeling was done on 72 experimental data, and the corrections of the obtained mathematical model was tested by (i) determining the value of R^2 and (ii) testing with 36 testing data. During this process, the "Wolfram Mathematica v.11" program was used in all the calculations.

Table 12.3 Testing experimental set for TIG welding operation [2].

	Welding Speed (cm/min)	Wire Speed (cm/ min)	Cleaning (%)	Gap (mm)	Current (A)	Front Height (mm)	Front Width (mm)	Back Height (mm)	Back Width (mm)
1	24	2	30	2.4	80	−0.066	6.123	0.801	5.541
2	24	2	30	3.2	80	0.114	5.979	0.682	4.633
3	24	2	70	2.4	80	−0.213	7.424	0.806	7.026
4	24	2	70	3.2	80	0.034	7.516	0.557	7.48
5	24	2	30	2.4	95	−0.167	8.481	0.713	8.34
6	24	2	30	3.2	95	−0.296	8.928	0.807	8.64
7	24	2	70	2.4	95	−0.219	9.677	0.688	9.717
8	24	2	70	3.2	95	−0.448	10.523	1.005	11.088
9	24	2	30	2.4	110	−0.281	10.871	0.713	11.142
10	24	2	30	3.2	110	−0.452	10.83	0.803	11.37
11	24	2	70	2.4	110	−0.651	13.986	1.09	14.146
12	24	2	70	3.2	110	−0.74	12.273	1.148	12.712
13	35	2	30	2.4	80	0.144	5.474	0.425	5.057
14	35	2	30	3.2	80	0.224	5.449	0.379	3.884
15	35	2	70	2.4	80	0.023	5.758	0.515	4.97
16	35	2	70	3.2	80	0.041	5.758	0.54	4.768
17	35	2	30	2.4	95	−0.094	6.665	0.613	6.304
18	35	2	30	3.2	95	−0.154	7.402	0.564	7.44
19	35	2	70	2.4	95	−0.179	7.614	0.61	7.557
20	35	2	70	3.2	95	−0.05	7.506	0.457	7.31
21	35	2	30	2.4	110	−0.433	8.011	0.868	8.047
22	35	2	30	3.2	110	−0.449	8.473	0.78	8.466
23	35	2	70	2.4	110	−0.396	9.652	0.782	10.277
24	35	2	70	3.2	110	−0.553	9.773	0.847	10.427
25	46	2	30	2.4	80	0.454	5.581	0.315	3.046
26	46	2	30	3.2	80	0.193	4.645	0.332	2.81
27	46	2	70	2.4	80	0.023	5.656	0.584	4.034

Table 12.3 contd. …

...Table 12.3 contd.

	Welding Speed (cm/min)	Wire Speed (cm/min)	Cleaning (%)	Gap (mm)	Current (A)	Front Height (mm)	Front Width (mm)	Back Height (mm)	Back Width (mm)
28	46	2	70	3.2	80	0.219	5.538	0.363	2.857
29	46	2	30	2.4	95	0.057	5.600	0.495	4.836
30	46	2	30	3.2	95	0.155	6.002	0.351	4.922
31	46	2	70	2.4	95	−0.189	5.859	0.729	5.201
32	46	2	70	3.2	95	−0.182	6.124	0.569	5.299
33	46	2	30	2.4	110	−0.368	6.927	0.748	6.775
34	46	2	30	3.2	110	−0.154	6.877	0.539	6.335
35	46	2	70	2.4	110	−0.35	7.63	0.65	7.869
36	46	2	70	3.2	110	−0.225	7.553	0.557	7.707

The general functional form of regression model used is as follows:

$$y = b_0 + \sum_1^5 b_i\, x_i + \sum_1^5 b_{ij}\, x_i\, x_j + \sum_1^5 b_{ijk} x_i\, x_j\, x_k + \sum_1^5 b_{ijkl}\, x_i\, x_j\, x_k\, x_l + \sum_1^5 b_{ijklm}\, x_i\, x_j\, x_k\, x_l$$
$$x_m + \sum_1^5 b_{ii}\, x_i^2 + \sum_1^5 b_{iiijj}\, x_i^3\, x_j^2 + \sum_1^5 b_{iijjk}\, x_i^2\, x_j^2\, x_k + \sum_1^5 b_{iijkl}\, x_i^2\, x_j\, x_k\, x_l + \sum_1^5 b_{iii}\, x_i^3 + \sum_1^5 b_{iiii}\, x_i^4 + \sum_1^5 b_{iiiii}\, x_i^5$$

$$(12.1)$$

where x, y, and b terms represent the coded values of the input variables, the response or output value and coefficient values determined by the least-squares method, respectively. The proposed mathematical model shows that the calculated values of R^2 are appropriate and since the differences between the measured values and the calculated values are

Table 12.4 Training and testing R^2 values of different models.

R^2	Output	Ref. [8]	4th Degree Polynomial Model [Present]	5th Degree Polynomial Model [Present]
Training	BH	0.789298	0.937399	0.964117
	BW	0.971766	0.990747	0.997344
	FH	0.847849	0.940308	0.97193
	FW	0.972834	0.990799	0.997724
Testing	BH	0.850297	0.961873	0.988406
	BW	0.982617	0.996965	0.997496
	FH	0.930275	0.973936	0.986197
	FW	0.970867	0.996155	0.999355

very small. The cross-section geometry of the weld seam accurately defines the engineering parameters.

Effect of TIG Welding Parameters on the Performance Factor

The change in the welding parameters also affects the dimensions of the section geometry of the weld seam. However, the maximum height (penetration) of the bottom height depends on all parameters. Jackson [15] defined welding technique performance factor (WTPF) which has a relationship amongst penetration, welding voltage, welding current and welding speed as follows

$$\text{WTPF} = \sqrt[3]{I^4 / SE^2} \tag{12.2}$$

where I, S, and E represents the source current (amperes), source rate (mm/min), and source voltage (volts), respectively. The performance factor of the welding technique for penetration is shown in Figure 12.2. It has been found that the welding technique reported by Jackson [15] has the same tendency as the performance factor results. Here, it is seen that the penetration increases with increasing welding technique performance factors.

Optimization of TIG Welding Parameters

The developed mathematical models are useful to choose the correct process parameters in order to obtain the desired weld seam quality or to estimate the weld seam quality for the given process parameters.

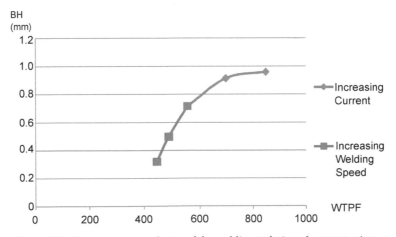

Figure 12.2 The performance factor of the welding technique for penetration.

These mathematical models facilitate process optimization. They also help (i) understanding the effect of process parameters to the weld seam quality and (ii) optimizing the weld seam quality in order to achieve a high-quality weld seam at a relatively low cost with high efficiency. As shown in Table 12.5, the top width, top height and bottom width have been minimized with four different optimization problems (Problems 1–4) in order to achieve the minimum weld seam cross-sectional area and maximum penetration. As a result of five different optimization studies (Problems 1–5), the minimum weld seam cross-sectional area with the maximum penetration condition was calculated. In Table 12.5, it can be seen that the smallest welding area is 7.0867 mm². However, maximum penetration cannot be achieved with the calculated input parameters when reaching this value. As a result of these calculations, the BH value is 0.06 mm. The obtained 0.06 mm value is not a desirable penetration value because such a small measurement value cannot be achieved in practice.

For this reason, the area of the welded suture geometry has been minimized and optimized for each case (5a–5f) by applying each constraint which is BH > 0.1, BH > 0.3, BH > 0.4, BH > 0.5, and BH > 0.7. The input parameters are placed into output functions BW, FH, and FW in order to determine welding area values during the optimization operation. In Table 12.5, it is demonstrated that BH value has the highest growth rate while the weld stitch cross-sectional area increased from 7.0867 mm² to 8.1101 mm².

Table 12.5 Optimum welding geometry measurements and input parameters based on DE [2].

Pr. No.	Optimization Problems	Welding Speed (cm/min)	Wire Speed (cm/min)	Gas Flow Rate (%)	Gap Distance (mm)	Welding Current (A)	Response	Test Number
1	FW Minimization	46	1.5	70	3.2	80	4.5552	TD-52
							4.423	–
2	FH Minimization	24	2	30	3.2	110	–0.8942	VD-12
							–0.74	–
3	BW Minimization	46	2	30	3.2	80	–0.0127	VD-25
							3.046	–
4	BH Maximization	24	2	70	3.2	110	1.1471	VD-12
							1.148	–
5	Weld Bead Sectional Area Minimization	46	2.5	30	2.4	80	7.0867	–

Besides that, for all welding area calculations for BH values were not as high as it was in the first condition. However, the amount of cross-sectional area of the weld seam increased. For this reason, the solution of problem number 5c with the input parameter values gives us the desired result at the point where the cross-sectional area of the weld seam is 8.1101 mm² (see Table 12.6).

Table 12.6 Optimization results for different BH constraints [1].

Optimization Problem	Area	BH (mm)	BW (mm)	FH (mm)	FW (mm)	a (cm/mm)	b (cm/mm)	c (%)	d (mm)	e (a)
								INPUTS		
5a	7.0867	0.067	1.298	0.395	4.746	46	2.5	30	2.4	80
5b	8.0234	0.25	2.806	0.207	4.766	46	1.5	30	3.2	80
5c	8.1101	0.448	3.328	0.0485	4.555	46	1.5	70	3.2	80
5d	10.6194	0.617	3.848	0.223	5.527	35	2.5	30	3.2	80
5e	14.1268	0.621	6.408	0.079	6.655	46	2.5	70	3.2	95
5f	14.1295	0.761	5.71	0.132	6.784	24	2.5	30	3.2	80

Conclusion

In this chapter, the effect of the Tungsten inert gas (TIG) welding process parameters (welding speed, welding current, gas flow rate, wire-speed, and gap distance) on the weld pool shape and the quality responses were investigated. The mathematical models have been developed for optimization and prediction of the weld pool geometry. A non-linear neuro-regression analysis has been applied to construct the relationship between welding process parameters and weld pool geometry in TIG welding. Mathematical calculations have been done in the Wolfram Mathematica v.11. In the optimization, the Differential Evolution method has been used. As a result of the work done, the optimization process has been carried out under the condition of the lowest possible value for the area of the weld bead geometry, which is the highest possible penetration amount. The improved mathematical model can precisely predict the weld seam cross-section geometry for TIG welding of 1100 grade aluminum plates of 1.6 mm thickness. The study reveals the importance of interactions in the input parameters and their effect on the weld seam cross-section geometry. The effect of welding parameters can be the same for various materials. The difference is only on source responses. Therefore, these parameters can also be prepared and tested on various material types in the future for weld seam cross-section geometry measurements.

References

[1] Turhan, F. 2017. 1100 serisi alüminyum malzemelerde tig kaynağı ile oluşan kaynak dikiş geometrisinin optimizasyonu. M.S. Thesis. Izmir Kâtip Celebi University, İzmir.

[2] Nanda, N.K. and Balasubramanian, K.R. 2011. Parametric optimization of tig welding on 316l austenitic stainless steel by grey-based Taguchi method. Adv. Mat. Res. 383-390: 4667–4671.

[3] Narang, H.K., Singh, U.P., Mahapatra, M.M. and Jha, P.K. 2011. Prediction of the weld pool geometry of tig arc welding by using fuzzy logic controller. IJEST 3(9): 77–85.

[4] Tarng,Y.S., Juang, S.C. and Lii,H.R. 1998. A comparison between the backpropagation and counter-propagation networks in the modeling of the TIG welding process. J. Mater. Process. Technol. 75: 54–62.

[5] Yang, L.J., Chandel, R.S. and Bibby, M.J. 1993. An analysis of curvilinear regression equations for modeling the submerged arc welding process. J. Mater. Process. Technol. 37(1-4): 601–611.

[6] Lee, J.I. and Rhee, S. 2000. Prediction of process parameters for gas metal arc welding by multiple regression analysis. Proc. Inst. Mech. E. Part B 214: 443–449.

[7] Kim, I.S., Son, J.S., Kim, I.G., Kim, J.Y. and Kim, O.S. 2003. A study on relationship between process variables and bead penetration for robotic arc welding. J. Mater. Process. Technol. 136: 139–145.

[8] Dutta, P. and Pratihar, D.K. 2007. Modeling of TIG welding process using conventional regression analysis and neural network-based approaches. J. Mater. Process. Technol. 184: 56–68.

[9] Joby, J. and Muthukumaran, S. 2015. Optimization of activated tig welding parameters for improving weld joint strength of aisi 4135 pm steel by genetic algorithm and simulated annealing. Int. J. Adv. Manuf. Technol. 6(72).

[10] Dey, V., Pratihar, D.K., Datta, G.L., Jha. M.N., Saha, T.K. and Bapat, A.V. 2010. Optimization and prediction of weldment profile in bead-on-plate welding of Al1100 plates using electron beam. Int. J. Adv. Manuf. Tech. 48: 513–528.

[11] Tarng, Y.S., Tsai, H.L. and Yeh, S.S. 1999. Modeling optimization and classification of weld quality in tungsten inert gas welding. Int. J. Mach. Tool. Manu. 39(9): 1427–1438.

[12] Esme, U. 2006. Effect of Pool Geometry on the Quality of Tig Welded Joints. PhD Thesis. Çukurova Üniversitesi. Fen Bilimleri Enstitüsü. Adana.

[13] Arpith, S., Singh, B.K. and Mastanaiah, P. 2017. Prediction and optimization of weld bead geometry for electron v beam welding of aisi 304 stainless steel. Int. J. Adv. Manuf. Tech. 89: 27–43.

[14] Gunaraj, V. and Murugan, N. 2000. Prediction and optimization of weld bead volume for the submerged arc process—Part 1. Weld. J. 286–294.

[15] Jackson, C.E. and Shrubsall, A.E. 1953. Control of penetration and melting ratio with welding technique. Weld J. 32(4): 172–178.

CHAPTER 13

Geometrical Optimization for a Cold Extrusion Process

Sezgin Yurtdas,[1] *Levent Aydin,*[2,]* *Cenk Kilicaslan*[1] and *H Irem Erten*[2]

Introduction

In fastening applications, the majority of fasteners are produced by using the cold forming procedure. These fasteners are generally generated in multi-stage processes by cold forging, pressure forging, and reducing or a combination of all procedures [1–3].

Cold forging, which is a type of metal forming process used in a broad spectrum of applications, from electronics to automotive and medical devices manufacturing, permits high-speed mass production with excellent mechanical properties in several applications. It is also proper for pieces which have excellent concentricity, narrow geometrical tolerances, smooth surface finish, and near net shape [1–3].

In the designing of the cold forging manufacturing processes, the forging process can be regarded as a complicated system which contains implicit interactions between the workpiece, forging force, temperature, forging die, tool and press influenced by environmental and tribological conditions. Therefore, it is a necessity to consider systematic design methods such as engineering optimization techniques. Engineering optimization is the matter which uses optimization techniques to obtain the best design and performance evaluation in engineering systems as quickly as possible. Hence, there are several optimization studies which

[1] Norm Cıvata San. ve Tic. A.Ş., A.O.S.B., İzmir, Turkey.
Emails: sezgin.yurtdas@normcivata.com; cenk.kilicaslan@normcivata.com
[2] İzmir Katip Çelebi University, Department of Mechanical Engineering, İzmir, Turkey.
Email: erteniremm@gmail.com
* Corresponding author: leventaydinn@gmail.com

are related to cold forging [1–3]. For instance, Han and Hua [4] studied about the contact pressure response in metal forming technology of cold rotary forging. They purposed using finite element (FE) methods to find out the contact pressure response in a complex cold rotary forging. For cold rotary forging, a 3D FE model is improved, and then, the contact pressure is calculated. As a result, the process parameters' effect on the contact pressure response is revealed. Krusic et al. [5] investigated the stochastic nature for the typical cold forging processes. They applied finite element computations on forwarding rod extrusion, closed-die forging, free upsetting, and retractile extrusion to examine the influence of Scatter's key process input parameters on the dimensional accuracy of products. The results show that for the four typical cold forging processes, stochastic interactions diverge and press stiffness has a significant effect on the stochastic relationships. Composite cold forging refers to the processing of hybrid raw parts at the same time by cold forging. Ossenkemper et al. [6] aimed to produce composite components by joining them with plastic deformation, and they manufactured composite shafts by forwarding rod extrusion of backward extruded steel cups. An analytical model has been improved to find the strength of the force of composite shafts which are generated by cold forging operations and push-out tests were done to identify the bond strength for the proposed model. Finally, they decided that calculated bond strengths were appropriate in accordance with the experimental values. In the world, approximate twenty different tribological tests have been suggested for the friction coefficients' empirical definition in the cold forging processes. Groche et al. [7] compared six test principles by using identical tribological systems, and they observed large differences between the empirically determined friction coefficients. For process parameters design, Zhu et al. [8] proposed a multi-objective optimization method by using the response surface methodology (RSM) approach and the Latin hypercube design method in order to check the product forming quality. As a result of this study, after conducting radial forging and experimental study, which is related to the mechanical property on the forming product, the feasibility of the multi-objective optimization method for product forming quality was validated. Francy et al. [9] studied on input process parameters (the shape of the workpiece; coefficient of friction, half die angle, logarithmic strain, die length and ram velocity) optimization in the extrusion process by Taguchi approach and DEFORM-3D software. The results showed that effective parameters have an important effect on decreasing the minimum extrusion force. Liang et al. [10] analyzed the serious earing defects in the initial extrusion process of an aluminum controller by using DEFORM-3D simulation. According to simulation results, the non-uniform velocity of metal flow is the basic cause of earing defects. Thus, a new process scheme which avoids the defects using a punch with a group of resistance ribs at the bottom was revealed. Results were compared with the initial process, and the earing defects have been saliently developed in the new process scheme. Forward and backward extrusion is important in the cold and hot forging industry and often used during the forging of complex parts. Hu et al. [11] developed a new method depend on a steady combined forward and backward extrusion tests in order to specify the friction factor for the cold forging process of the complex parts. FE simulations optimized important parameters that define the deformation degree of the forward and backward extrusion and the punch geometry and die in order to develop the sensitivity to friction. As a result, according to simulation results, they constructed two groups of calibration curves for different ranges of friction factors. Jin et al. [12] proposed an optimization method for the cold orbital forging of the component with a deep and narrow groove. As a beginning, three types of metal flow mode are offered based on the preform and tool geometries, and then they analyzed the cold orbital forging processes for three types of metal flow mode using finite-element simulations and forging defects of underfilling, the folding and poor stress state of the tools are found out. Lilleby et al. [13] studied the cold pressure welding of seriously deformed aluminum. Firstly, pure aluminum is exposed to single-pass equal channel angular pressing and then assembled by divergent extrusion. The mechanical integrity of the spliced components is documented by using notch tensile testing and it showed that the strength and the flow properties of the SPD processed base material are continued across the joint after cold welding.

In this chapter, extrusion die geometry which is prepared for bolt production is considered, and for this die, the die design conditions that provide minimum extrusion force requirement was determined. For this reason, the primary objective is to obtain the minimum extrusion force by establishing the relationship between the inputs as three main parameters (extrusion angle, temperature and friction coefficient) and the output (extrusion force) in the extrusion die to form the workpiece. In light of these objectives, numerical analyses were performed in simufact.forming a finite element software program for the determined parameters, and then the data was obtained. Then, different regression models were created through a commercial software Mathematica based on neuro-regression approach. Besides this, the accuracy of the constructed models has been checked through $R^2_{training}$ and $R^2_{testing}$ values. After testing the reliability of the models, and to optimize the input parameters that minimize the output also, stochastic methods, which are Differential Evolution (DE), Random Search (RS) and Simulated Annealing (SA), and a deterministic one Nelder-Mead were used in Mathematica.

Method

Numerical models of cold extrusion operations were prepared in simufact. forming finite element software. Mechanical models were also coupled with thermal analysis to consider temperature effects on the flow stress of the workpiece material. Schematic representation of numerical models is shown in Figure 13.1. The extrusion process was simulated with 2D models due to the axisymmetry. As shown in the figure, numerical models consist of stationary and moving dies and the workpiece. Dies were modeled as rigid. Workpiece material, medium carbon steel, was modeled as plastic material and true stress-true plastic strain curves between temperatures of 20°C and 400°C and strain rates between 1 and 50 s^{-1} was defined to the software. In 2D models, 3.444 quad elements were used in the mesh of the workpiece.

Problem Description

In this study, the extrusion die has been taken into consideration for the bolts. Factors which affect the operation of the die are extrusion angle, temperature, and friction coefficient (see Figure 13.2). Three steps have

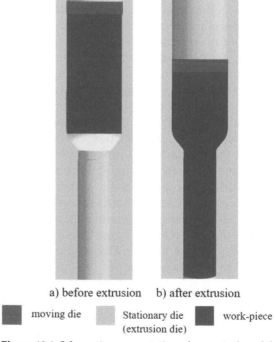

a) before extrusion   b) after extrusion

■ moving die   ▨ Stationary die   ■ work-piece
(extrusion die)

Figure 13.1 Schematic representation of numerical model.

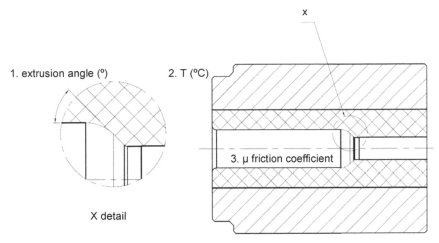

Figure 13.2 Technical drawing of the extrusion die.

been performed for the design process: (i) the data for the determined parameters were obtained by performing numerical analyses in simufact. forming software program (ii) the different regression models were created and (iii) optimization studies were carried out in Mathematica software.

Multiple nonlinear regression analysis was used in the modeling step. By the neuro-regression approach, firstly, 30 simulation results were randomly divided into training and testing data for forging force. The training data consists of 80% of all data randomly selected, and it is necessary to check the accuracy of the constructed model while the testing data is obtained by 20% of all data (see Tables 13.1 and 13.2). After the separation of the data, five different regression models as the general functional form (see Table 13.3) have been introduced for the engineering parameters and $R^2_{training}$, and $R^2_{testing}$ values for each model have also been calculated by Mathematica (see Table 13.4). In these models, the input parameters X_1, X_2, and X_3 represent the extrusion angle, temperature, and friction coefficient, respectively.

Results and Discussion

In this part, the maximum and minimum values of each model were calculated, and then their compatibility with the maximum and minimum values in the actual data was checked (see Table 13.5). Among the five models introduced in Table 13.4, model 3, which has the highest testing value and the closest maximum and minimum values to actual data, was chosen as the candidate model for the optimization. Then, the selected engineering model is investigated by introducing three optimization problems which

Table 13.1 Training simulation data.

Case	Extrusion Angle X_1 [°]	Temperature X_2 [°C]	Friction Coefficient X_3	Extrusion Force [t]
1	40	100	0.09	11.6481
2	35	25	0.18	26.3745
3	40	25	0.18	26.5189
4	35	100	0.12	15.465
5	40	25	0.05	8.03995
6	45	25	0.09	12.6291
7	45	100	0.12	15.8635
8	25	250	0.12	15.0349
9	30	100	0.18	24.9746
10	25	250	0.18	24.2671
11	30	25	0.18	26.3603
12	45	100	0.05	7.83371
13	30	250	0.12	14.8395
14	30	25	0.05	7.67811
15	35	250	0.12	15.1231
16	45	100	0.18	25.2247
17	25	100	0.05	7.43887
18	40	100	0.12	15.5502
19	25	25	0.05	7.80922
20	30	100	0.12	15.0971
21	40	25	0.12	16.4658
22	25	25	0.09	11.8979
23	35	25	0.05	7.99821
24	35	25	0.09	11.5575

Table 13.2 Testing simulation data.

Case	Extrusion Angle X_1 [°]	Temperature X_2 [°C]	Friction Coefficient X_3	Extrusion Force [t]
25	40	250	0.18	24.0229
26	45	25	0.18	26.6905
27	35	250	0.18	24.0798
28	25	25	0.12	15.8995
29	35	100	0.05	7.5532
30	35	100	0.09	11.5221

Table 13.3 General functional form for extrusion force.

Model No.	Models
1	$$\sum_{n=0}^{3} a_n(1+X_1+X_2+X_3)$$
2	$$\sum_{n=0}^{3} \frac{a_n(1+X_1+X_2+X_3)^1}{b_n(1+X_1+X_2+X_3)^1}$$
3	$$\sum_{n=0}^{3} a_n(1+X_1+X_2+X_3)^2$$
4	$$\sum_{n=0}^{3} \frac{a_n(1+X_1+X_2+X_3)^2}{b_n(1+X_1+X_2+X_3)^2}$$
5	$$\sum_{n=1}^{3} 1 + Sin(X_1) + Sin(X_2) + Sin(X_3) + 1 + Cos(X_1) + Cos(X_2) + Cos(X_3)$$

include minimizing extrusion force with appropriate constraints. The aim of problem 1 is to mathematically see the limitation of result values of the problem while design variables are chosen in continuous domains. In problem 2, while the extrusion angle and temperature are analyzed in the continuous domain, it has taken the available values given in the simulation data for friction coefficient. In this way, the dependence of the friction coefficient parameter on the selection domain was evaluated for optimization. In problem 3, all design variables have been created considering the actual working conditions in the industry. As shown in Table 13.6, for the problems, extrusion force is minimized based on the DE, RS, SA, and NM methods. When the first two problems were considered, the same optimum values were obtained regardless of whether they were stochastic or deterministic. So that, $X_1 = 27.7653$, $X_2 = 116.1268$, $X_3 = 0.05$, and forging force = 7.3422. In problem 3, the optimum results are $X_1 = 30$, $X_2 = 100$, $X_3 = 0.05$, and forging force = 7.3673. These values are higher than optimum values as expected. However, it was decided that there is approximately 0.31% tolerable difference for the industry conditions. This showed that it is not necessary to make the working conditions of the current process more sensitive. In other words, it can be approached the optimum solution in a slight difference with the current sensitivity.

Table 13.4 Models for extrusion force.

Model No.	Model	R² Training	R² Testing
1	$0.65727 - 0.00490\,X_1 - 0.0095\,X_2 + 141.0383\,X_3$	0.9861	0.9832
2	$\dfrac{880.8394 + 11.0304\,X_1 + 0.4689\,X_2 + 23257.2733\,X_3}{333.4622 + 0.4202\,X_1 + 0.1041\,X_2 - 792.6936\,X_3}$	0.9984	0.9987
3	$5.4629 - 0.09870\,X_1 + 0.0020\,X_2^2 - 0.0035\,X_2 - 0.00006\,X_1\,X_2 + 0.00004\,X_2^2 + 55.5916\,X_3 - 0.1065\,X_1\,X_3 - 0.08841\,X_2\,X_3 + 402.62914\,X_3^2$	0.9994	0.9998
4	$\dfrac{\begin{aligned}&(1153.551 - 271.8023X_1 + 5.4732\,X_1^2 + \\ &16.008\,X_2 - 0.2322\,X_1\,X_2 + 0.0524X_2^2 \\ &+ 76803.8248\,X_3 - 546.7535\,X_1\,X_3 + \\ &544.9389\,X_2\,X_3 - 1664715.6592\,X_3^2)\end{aligned}}{\begin{aligned}&(2413.1719 - 16.4347X_1 + 0.3411X_1^2 \\ &+ 7.5118X_2 - 0.0199X_1X_2 + 0.005X_2^2 \\ &- 28408.8632X_3 - 32.6094X_1X_3 - \\ &21.8497X_2X_3 + 54583)\end{aligned}}$	0.9999	0.9488
5	$773.343 - 0.0664\,\mathrm{Cos}(X_1) - 0.844\mathrm{Cos}(X_2) - 767.7328\,\mathrm{Cos}(X_3) + 0.2521\,\mathrm{Sin}(X_1) + 2.2158\,\mathrm{Sin}(X_2) + 51.6142\,\mathrm{Sin}(X_3)$	0.9983	0.9930

Table 13.5 Maximum and minimum values for extrusion force.

Model	Minimum Value [t]	Maximum Value [t]
1	5.0955	25.6823
2	7.3757	26.2626
3	7.3442	26.3221
4	7.2538	The function is unbounded
5	7.7589	29.8866

Table 13.6 Optimization problems and optimum results.

			Design Variables			
Problem No.	Problem Definition	Method	X_1 [°]	X_2 [°C]	X_3	Extrusion Force [t]
1	Minimize Extrusion Force (X_1, X_2, X_3) Subjected to $25 < X_1 < 45$, $25 < X_2 < 250$, $0.05 < X_3 < 0.18$	DE	27.7653	116.1268	0.05	7.3442
		RS	27.7656	116.1258	0.05	7.3442
		SA	27.7655	116.1257	0.05	7.3442
		NM	27.7653	116.1256	0.05	7.3442
2	Minimize Extrusion Force (X_1,X_2,X_3) Subjected to $25 < X_1 < 45$, $25 < X_2 < 250$, $X_3 \in \{0.05,0.09,0.12,0.18\}$	DE	27.7653	116.1256	0.05	7.3442
		RS	27.7653	116.1256	0.05	7.3442
		SA	27.7653	116.1256	0.05	7.3442
		NM	27.7653	116.1256	0.05	7.3442
3	Minimize Extrusion force (X_1,X_2,X_3) Subjected to $X_1 \in \{25,30,35,40,45\}$, $X_2 \in \{25,100,250\}$, $X_3 \in \{0.05,0.09,0.12,0.18\}$	DE	30	100	0.05	7.3673
		RS	30	100	0.05	7.3673
		SA	30	100	0.05	7.3673
		NM	30	100	0.05	7.3673

Conclusion

In the present study, simulation and optimization studies on the extrusion die geometry which is prepared for bolt production were carried out to investigate the effects of the input parameters (extrusion angle, temperature and friction coefficient) on the extrusion force (output). The main aim is to provide the minimum extrusion force with valid optimum parameters in the extrusion die to form the workpiece. Therefore, numerical analyzes were applied using simufact.forming a finite element software program and then data was obtained for each operating parameter and extrusion force. We consider the neuro-regression approach and developed five regression models by a commercial software Mathematica. According to the accuracy of the model, the most appropriate model was selected. In the optimization step for an appropriate model, the calculations have been performed based on DE, RS, SA, and NM to optimize the input parameters that minimize the extrusion force. As a conclusion, the optimum solution was achieved with a slight difference from industry conditions with the current sensitivity thanks to the constructed regression model for the problem.

References

[1] Kroiß, T. and Engel, U. 2013. Optimization of Tool and Process Design for the Cold Forging of Net-Shape Parts by Simulation, Process Machine Interactions, Springer, Berlin, Heidelberg, pp. 419–437.

[2] Groche, P. 2018. Friction Coefficients in Cold Forging: A Global Perspective. 261–264.

[3] Kilicaslan, C. and Ince, U. 2017. Civata Soguk Dovme Isleminde Kalip Omrunun Arttirilmasi: Dovme Kademe Tasariminin Etkisi. SAUJS 21: 961–967.

[4] Han, X. and Hua, L. 2013. 3D FE modelling of contact pressure response in cold rotary forging. Tribol. Int. 57: 115–123.

[5] Krusica, V., Masera, S., Pristovsek, A. and Rodica, T. 2009. Adjustment of stochastic response pf typical cold forging systems. J. Mater. Process. Technol. 209: 4983–4993.

[6] Ossenkemper, S., Dahnke, C. and Tekkaya, A.E. 2019. Analytical and Experimental Bond Strength Investigation of Cold Forged Composite Shafts. 264: 190–199.

[7] Groche, P., Kramer, P., Bay, N., Christiansen, P., Dubar, L., Hayakawa, K., Hue, C., Kitamura, K. and Moreau, P. 2018. CIRP Annals-Manufacturing Technology 67: 261–263.

[8] Zhu, F., Wang, Z. and Lv, M. 2016. Multi-objective optimization method of precision forging process parameters to control the forming quality. Int. J. Adv. Manuf. Tech. 83: 1763–1771.

[9] Francy, K.A., SrinivasaRao, D.C. and Gopalakrishnaiah, P. 2019. Optimization of direct extrusion process on 16MnCr5 and AISI1010 using DEFORM-3D. Procedia Manuf. 30: 498–505.

[10] Cheng-liang, H., Li-fen, M., Zhen, Z., Bing, G. and Bing, C. 2012. Optimization for extrusion process of aluminum controller housing. T. Nonferr. Metal. Soc. 22: 48–53.

[11] Hu, C., Yin, Q. and Zhao, Z. 2017. A novel method for determining friction in cold forging of complex parts using a steady combined forward and backward extrusion test. J. Mater. Process. Technol.

[12] Jin, Q., Han, X., Hua, L., Zhuang, W. and Feng, W. 2018. Process optimization method for cold orbital forging of component with deep and narrow groove. J. Manuf. Process. 33: 161–174.

[13] Lilleby, A., Grong, Ø. and Hemmer, H. 2010. Cold pressure welding of severely plastically deformed aluminium by divergent extrusion. Mat. Sci. Eng. A-Struct. 527: 1351–1360.

CHAPTER 14

Aerodynamic Optimization of a Compressor Rotor Using Genetic Algorithm

Orçun Kor[1,]* and *Sercan Acarer*[2]

Introduction

The internal flow inside a turbo engine can be considered one of the most challenging fluid mechanics problems. The flow of air through the blades in the turbo engine is affected by many parameters and constraints, which makes it a very complicated process, and therefore numerous objectives need to be considered simultaneously [1, 2]. Rotating and stationary bodies inside the turbo engine create transonic flow with vortices that may cause catastrophic failures, and even the loss of the whole engine. A compressor designer, for instance, is expected to control such phenomena smartly while delivering requirements such as pressure ratio, mass flow rate and efficiency at a certain point, providing that the component would still be able to operate at off-design and extreme conditions. Thus, the designer would provide the requirements and also achieve a satisfying stall margin, while keeping his design as uncomplicated as possible with the final product having a minimum weight.

[1] Compressor Aerodynamicist, Technical Leader at TUSAŞ Engine Industries, İstanbul Teknoloji Geliştirme Bölgesi, Sanayii Mah. Teknopark Bulvarı, No: 1/7A 7301, İstanbul, Turkey.
[2] Assistant Professor in Mechanical Engineering Department of Izmir Katip Çelebi University, Izmir Katip Çelebi Universitesi, Makina Mühendisliği Bölümü Balatçık Kampüsü, Çiğli, İzmir.
Email: sercan.acarer@ikc.edu.tr
* Corresponding author: orcun.kor@tei.com.tr

Satisfying such challenging requirements, which conflict with each other, by human effort is possible, as has been done for decades. However, remarkable advancements in the turbomachinery world have already been achieved through the years. For further technological advancements, human efforts may be insufficient [3]. Moreover, even for an over-productive engineer, these efforts would be time-consuming, and the designs are prone to human-sourced errors, as this situation rises, so does the need for the use of automatic optimization techniques. As pay-back, the use of computational methods and algorithms becomes indispensable [4].

Many optimization techniques have been developed over the years, and each of them has its own superiority amongst others. One who has a promising initial design may rely on gradient-based techniques since the actual design is similar to the initial one, and the process is less vulnerable to the threat of sticking into the local optimum. The method, by definition, has its limits, where it is mathematically applicable only if the function which is trying to be optimized is differentiable, convex, and continuous [5]. The designer obtains the optimum quickly and time-effectively by using gradient methods. Nevertheless, in case that the design variables are allowed to take discrete values, differentiation problems will likely to occur. Even if the objective functions are differentiable, then the complex, discrete, and usually non-differentiable nature of the turbomachinery flows limits the abilities of gradient-based approaches. Novel gradient-based methods such as adjoint operator and control theory are reported to unravel the mathematical barriers and computational disadvantages that are suffered in the use of classical gradient-based methods. This method uses the objective function sensitivities for the design variables, where its superiority is its ability to exploit local information during the solution refinement. This way, the initial points do not lead the optimizer to a local optimum [6].

Besides that, there are stochastic methods, which refer to a collection of methods for finding the optimum when randomness is present [7]. Evolutionary algorithms (EA) among these methods are popular since they are easy to implement. The methods can be coupled with other design/analysis tools since the gradient information is not utilized [8]. The analytical limit for non-differentiable, discrete functions, are not a problem for EAs, which makes them preferable in optimization tasks, where the search space has singularities or other mathematically non-definable locations. Additionally, this method can find the global optimum regardless of the initial guess. The drawback is the slow convergence rate of EAs methods, since finding the global optimum necessitates a large number of function evaluations. Moreover, detecting the global optimum can require many iterations, especially in large design spaces. The disadvantages, however, are overcome by either using meta-models or making use of lower-order objective functions, and hybridizing gradient-based and evolutionary algorithms [9, 10, 11].

The Problem

As mentioned above, turbomachinery aerodynamics has a complex nature, where 3D flow structures, separation regions, and backflows play essential roles. The problem at hand is multi-modal and discontinuous; thus, there is a risk of sticking to a local minimum or running into an infeasible design for which the flow simulation does not converge when the gradient-based numerical optimization algorithm is used [12]. Therefore, a well-known EA method, namely, Genetic Algorithm (GA), is used in the scope of this work.

In general, an optimization problem can be defined as [13]:

Minimize

$$f_i(\bar{x}) = 0 \tag{14.1}$$

Subject to

$$g_j(\bar{x}) \leq 0, j = 1 \ldots m \tag{14.2}$$

$$h_k(\bar{x}) = 0, \ k = 1 \ldots n \tag{14.3}$$

$$x_p^{low}(\bar{x}) \leq x_p \leq x_p^{up}(\bar{x}), p = 1 \ldots q \tag{14.4}$$

with

$$\bar{x} = \begin{Bmatrix} x_1 \\ x_2 \\ x_3 \\ \ldots \\ x_q \end{Bmatrix} \tag{14.5}$$

where, $f_i(\bar{x})$ is objective function, $g_j(\bar{x})$ is inequality constraints, $h_k(\bar{x})$ is equality constraints and $\bar{x}$ is design variable vector.

The penalty function method is used in constraint handling, where designs violating constraints (i.e., $g_j(\bar{x}) \leq 0$) are penalized such that their objective function values are worsened virtually. G_j is evaluated as zero if there is no constraint violation regarding Equations 14.6 and 14.7.

In case a constraint is violated, a penalty term is added to the constraint function value as given in Equations 14.6 and 14.7. The penalty multiplier, r_k, is increased in successive iterations to make the optimizer less flexible against violations towards the last generations of the optimization process [14].

$$\phi_i = \phi(\bar{x}, r_k) = f_i(\bar{x}) + r_k \sum_{j=1}^{m} G_j [g_j(\bar{x})] \tag{14.6}$$

$$G_j = \{max[0, g_j(\bar{x})]\}^2 \tag{14.7}$$

In this study, the aerodynamic optimization of a 1 stage rotor blade inspired by NASA's Rotor37 is investigated. The aim of the optimization process to minimize the diffusion factor (DF) while keeping pressure ratio (PR$_{tt}$), mass flow rate, and tip camber at pre-determined values, as tabulated in Table 14.1. Although the problem seems to have a single objective, the definition of the diffusion factor itself provides a multi-dimensionality to the problem. It represents the diffusion from leading to the trailing edge of the blade as [15]:

$$DF = 1 - \frac{w_2}{w_1} + \left(\frac{\Delta V_\theta}{2\sigma w_1} \right) \tag{14.8}$$

Diffusion factor is a measure of static pressure rise and relative velocity deceleration in the blade passage. In Equation 14.8, w_1 and w_2 corresponds to inlet and blade exit relative velocity magnitudes at each blade height location, respectively. ΔV_θ is the absolute (stationary frame) rotational (tangential) velocity rise in the rotor at each blade height position. The parameter σ represents the solidity which is defined as the blade chord length to pitch the (tangential) distance between the two blades at each blade height position. DF values at 11 blade height positions (from root to tip) are averaged along the blade height to give a measure of the mean diffusion factor for the blade. Diffusion factor indirectly drives the values of efficiency and stall margin of a compression system [15]. Stall margin can be defined in terms of pressure ratio and mass flow rate as:

$$SM = \left[\frac{PR_{tt-stall}}{PR_{tt-design}} \times \frac{\dot{m}_{design}}{\dot{m}_{stall}} - 1 \right] \times 100 \tag{14.9}$$

where $\dot{m}_{design}$ and $\dot{m}_{stall}$ refers to the mass flow rate passing through the compressor in the design and stall (minimum mass flow) operational conditions. PR$_{tt}$ refers to the ratio of blade exit to inlet total pressures. Similar to DF, it is also averaged along the blade height to give an average value for the blade.

Figure 14.1 represents the compressor geometry in the axial-symmetric (meridional) plane. The design variables are rotor (blade) revolutions per minute (RPM), radial positions R1 and R2 at the blade inlet, and radial distribution of blade exit angle, which is the angle between the Z axis and the flow vector relative to the rotating blade.

Table 14.1 presents the objective and the constraints for the design variables. 'Camber' refers to β angle difference between the inlet and exit of the blade, therefore the turning imposed by the blade.

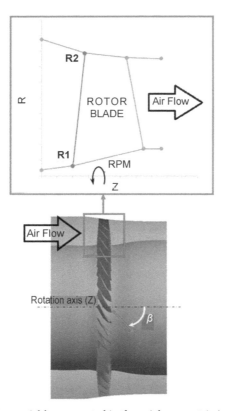

Figure 14.1 Design variables presented in the axial-symmetric (meridional) plane.

Table 14.1 Objective and constraints.

Minimize	$$DF = 1 - \frac{w_2}{w_1} + \left(\frac{\Delta V_\theta}{2\sigma w_1} \right)$$
Subject to	$1.96 < PR_{t-t}; \ PR_{t-t} < 2.10$
	$Camber_{tip} < 13.5$
	$\dot{m} = 20.19 \ kg/s$
Where	$\bar{x} = \{RPM \ \ R_1 \ \ R_2 \ \ \theta_1 \ \ \theta_2 ... \theta_{11}\}$

Fluid-flow Models

The governing equation for the fluid motion in a continuum is the Navier-Stokes Equations. For complex turbulence flow, to reduce the computational time, instead of solving all the length and time scales within the flow, time-varying velocities of any direction 'i' are decomposed into low frequency (mean) and high frequency (fluctuating) components such that:

$$v_i = \overline{v}_i + v_i'$$ (14.10)

Substituting these components into the original Navier-Stokes Equations and dropping the bar in Equation 10 that indicates time-average velocity yields to Reynolds-Averaged Navier-Stokes (RANS). Therefore, Navier-Stokes Equations for a compressible Newtonian fluid (i.e., air) can be written as [16]:

$$\frac{\partial \rho v_i}{\partial t} + \frac{\partial}{\partial x_j}\left(\rho v_i v_j\right) = -\frac{\partial p}{\partial x_i} + \frac{\partial}{\partial x_j}\left[\mu\left(\frac{\partial v_i}{\partial x_j} + \frac{\partial v_j}{\partial x_i}\right) - \overline{\rho v_i' v_j'}\right]$$ (14.11)

where the term $\overline{\rho v_i v_j}$ is the so-called Reynolds stresses and takes into account mixing due to turbulent fluctuations, p is pressure, ρ is density, μ is dynamic molecular viscosity.

The mass continuity equation is unaffected by this decomposition, and it can be written as:

$$\frac{\partial \rho}{\partial t} + \frac{\partial}{\partial x_j}(\rho v_j) = 0$$ (14.12)

Finally, the energy equation can be written as:

$$\frac{\partial \rho H}{\partial t} - \frac{\partial p}{\partial t} + \frac{\partial}{\partial x_j}\left(\rho v_j H\right) = \frac{\partial}{\partial x_j}\left(k\frac{\partial T}{\partial x_j} - \overline{\rho v_j' h}\right) + \frac{\partial}{\partial x_j}\left\{v_j\left[\mu\left(\frac{\partial v_i}{\partial x_j} + \frac{\partial v_j}{\partial x_i}\right) - \overline{\rho v_i' v_j'}\right]\right\}$$ (14.13)

where H is total enthalpy which includes the contribution from turbulent kinetic energy, h is enthalpy and k is thermal conductivity. The term $-\overline{\rho v_j' h}$ takes into account the additional flux due to turbulence, whereas the term containing Reynolds stresses considers viscous dissipation due to both mean and fluctuating velocity components.

For the turbulent terms, the well-known Boussinesq turbulent viscosity assumption is used to simulate Reynolds stresses, where the molecular viscosity is replaced by an effective (molecular+turbulent) viscosity field to take into account the effects of turbulent fluctuations in an isotropic manner, that is:

$$-\overline{\rho v_j' v_j'} = \mu_{turbulent}\left(\frac{\partial v_i}{\partial x_j} + \frac{\partial v_j}{\partial x_i}\right)$$ (14.14)

Similarly, turbulent conductivity ($\mu_{turbulent}/Pr_{turbulent}$) is defined as:

$$\overline{\rho v_j' h} = \frac{\mu_{turbulent}}{Pr_{turbulent}} \frac{\partial h}{\partial x_j} \tag{14.15}$$

The problem is now reduced to obtaining $\mu_{turbulent}$ as a function of space. Shear Stress Transport (SST) turbulence model with intermittency and cross-flow transition sub-models. ANSYS CFX reattachment modification sub-model [17] is selected to model the turbulent viscosity field. ANSYS CFX 19.1 is utilized to solve these equations iteratively for the considered compressor flow-fields by the vertex-centered finite volume method [17]. Such a numerical solution is called Computational Fluid Dynamics (CFD). SST models are known to provide robust results in flow fields with adverse pressure gradients [18]. Moreover, validation by the authors is performed as in the open test case of NASA Rotor 37 [19].

The fluid-flow models described so far require substantial computational power such that 1.3×10^6 finite control volumes are used to solve the compressor flow-field, and a single simulation converges in one day by using Intel Xeon E5-2680 V4 processors with 2.4 GHz clock speed. Therefore, such CFD modeling is not eligible for optimization studies where a vast number of simulations are performed. Throughflow models are reduced-order models for turbomachinery flows such that only the flow-fields in the meridional plane (axial-radial) are solved with swirl (tangential) flow. In these models, the unresolved effects (the effect of the blades) are considered by the integral momentum approach and empirical correlations for pressure losses and flow guidance by the blades.

The fidelity of the turbomachinery design increases with the complexity of the flow modeling, where such complexity is associated with a higher computational cost and run time [20] which makes the use of fast and reliable pre-3D design methods (a.k.a. throughflow methods) inevitable.

Streamline curvature (SLC) method is one of the most common and reliable solution methods for throughflow calculations. Therefore, it is utilized as a two-dimensional design tool. This method plays a crucial role in the turbomachinery aerodynamic design. Such methods have their limits in physical modeling; however, they are useful in the initial stages of design [21]. In this method, the flow equations described in Equation (16) for the meridional (axially symmetric) plane are cast in streamline coordinates, assuming meridional viscous effects are negligible.

$$V_m \frac{\partial V_m}{\partial q} = \frac{\partial H}{\partial q} - T\frac{\partial s}{\partial q} - \frac{V_\theta}{r}\frac{\partial(rV_\theta)}{\partial q} + V_m \sin\varepsilon \frac{\partial V_m}{\partial m} + K_m V_m^2 \cos\varepsilon - F_q \tag{14.16}$$

The computational domain is established by the number of nodes defined by intersections of streamlines with the predefined quasi-orthogonals (QO), which are lines roughly normal to the streamlines. During the simulations, nodes float on the QOs to iteratively improve the shape of the streamlines to satisfy mass flow continuity. The resulting simulations are around three orders of magnitude faster relative to 3D CFD and converge within seconds. Such modeling is described by Cumpsty [22] and Acarer and Özkol [23] in detail. The computational domain is the meridional plane as depicted in Figure 14.1. Axstream Software by SoftInWay is used [24].

Optimizer Configurations

In this study, a conventional type of genetic algorithm methodology which works with binary representation is used. The details of the algorithm can be found in the previous work of one of the authors [4]. An in-house optimization program developed by authors' is used for this issue, which utilizes the algorithm summarized in Figure 14.2.

The initial population fed to the algorithm consists of 40 randomly generated individuals. Among these, the design which provides the lowest objective function value is picked as "initial best" and the objective function improvement obtained during optimization is evaluated with reference to this "initial best" design. The percentage of improvement can be calculated as:

$$\% \ improvement = \frac{f_{optimum} - f_{initial \ best}}{f_{optimum}} \times 100 \qquad (14.17)$$

Reproduction step is performed via the tournament method, and the mutation ratio is picked as 1%. The optimizer stops the search once a given number of generations are achieved, which is decided by the designer. The maximum number of generations is taken as 15.

Optimization Results

Optimization is stopped at the end of 15 generations, which corresponds to 72 hours of computational time. It is observed that a convergence has been reached at the end of the 5th generation (~ 24 hours computational time), which points to the situation that the initial guess is already close to the optimum. The optimization is, however, not stopped, since the mutation process may still end up with a better solution. Nevertheless, no objective function improvement is observed after the 5th generation.

Note that objective function values at each generation are normalized with reference to the best design obtained in the initial family. The

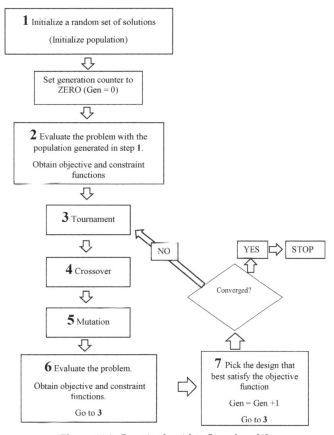

Figure 14.2 Genetic algorithm flow chart [4].

relative improvement of the objective function is assessed with reference to the initial family's best member. The normalized objective function is calculated as:

$$\bar{f} = \frac{f_{best,gen}}{f_{best,init}} \tag{14.18}$$

Flow path comparison of baseline and optimum flow path geometries are shown in Figure 14.4. Exit β angle distribution of baseline and optimum blade geometries are shown in Figure 14.5. At the end of the optimization process, an improvement of 3.5% regarding the objective function is reached as seen in Figure 14.3, while all constraints are satisfied. Even though the geometries compared in Figures 14.4 and 14.5 are very similar to each other, the optimizer is able to achieve a small but important adjustment on the design, which may not be captured manually by even

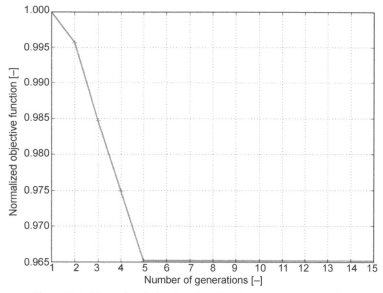

Figure 14.3 Normalized objective function for successive generations.

Table 14.2 Optimization results.

	Baseline	Optimum
DF	0.59	0.57
$Camper_{tip}$	9.47	11.80
$\dot{m}\,(kg/s)$	20.19	20.19
RPM	17400	17975

an experienced designer. The final touch made by the genetic algorithm on the flow path and blade exit β angle distribution is seen to provide a significant improvement on diffusion factor, due to the well-known design rule that a higher diffusion factor results in higher wake momentum thickness, therefore higher losses, as suggested by Lieblein [25]. Excessive values of diffusion factor also can be considered as a blade stall inception condition [15]. The numerical values of objective and constraint functions are given in Table 14.2.

As discussed above, optimization based on lower-order methods provides the designer with a fast and robust prediction on improving the aerodynamic design. However, the design should be finalized by using reliable 3D CFD calculations to assess the design thoroughly. Therefore, CFD calculations of the baseline and optimum geometries have been performed and explained in the following section.

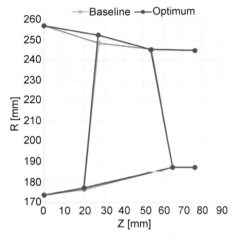

Figure 14.4 Baseline and optimum flow geometries.

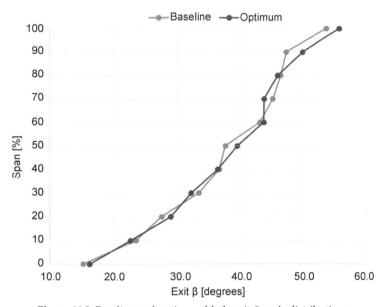

Figure 14.5 Baseline and optimum blade exit β angle distributions.

Assessment of Optimization Results in the CFD Model

The CFD model for the baseline and optimized geometries are presented in Figure 14.6. Notable differences such as a higher setting angle towards the rotor tip are observed. Thickness is kept constant for a consistent comparison. Since the geometry is periodic in the rotational direction,

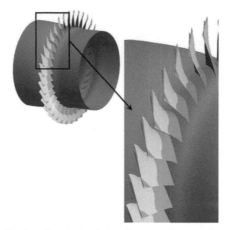

Figure 14.6 Baseline (blue) and optimized (gray) rotors. Orange flow path at the blade root is for the baseline rotor.

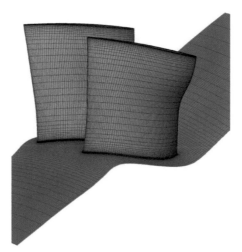

Figure 14.7 Periodic computational mesh for the baseline rotor considering.

only a single rotor is modeled, and periodic boundary conditions are implemented. The computational mesh (with 1.3×10^6 finite control volumes) for the baseline geometry is presented in Figure 14.7. The mesh is duplicated on the periodic surfaces for convenience. A similar mesh is also created for the optimized geometry.

The global performances of the baseline and optimized geometries are presented in Table 14.3, Figures 14.8 and 14.9. Due to the fidelity difference between the more accurate 3D CFD and the less accurate 2D throughflow simulations, predicted performances might differ. However, the CFD

Table 14.3 Global performances predicted by 3D CFD simulations ($PR_{tt} = 1.97$).

	Baseline	**Optimum**	**Improvement (%)**
DF	0.58	0.55	4.01
SM (%)	4.68	11.60	6.20
$\dot{m}$ (kg/s)	19.50	20.19	–
η_{is} (%)	81.5	80.3	–1.20
RPM	17400	17975	

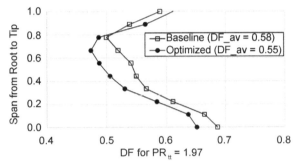

Figure 14.8 Diffusion factor distribution of baseline and optimized geometries (from root to tip).

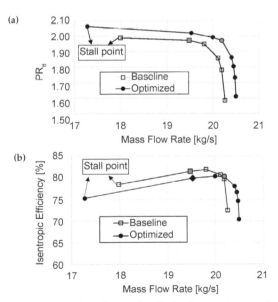

Figure 14.9 Performance maps of baseline and optimized geometries (red symbols refers to $PR_{tt} = 1.97$ operational conditions of baseline and optimized geometries) (a) PR_{tt} (b) Isentropic efficiency.

results presented in Table 14.3, Figures 14.8 and 14.9 reveal the intended improvement relative to the baseline geometry. It can be seen from the results that the baseline geometry inspired from Rotor 37 is not capable of providing the desired mass flow rate and total pressure ratio (ratios of exit to inlet total pressures) at the same time.

On the other hand, optimum geometry offers a less of a diffusion factor with respect to baseline, which in turn results in a higher stall margin expectedly. Please note that comparison of baseline and optimum geometries are performed with reference to the operation conditions (also called 'points') which satisfy the same pressure ratio (PR_{tt}) of 1.97. An undesired drop in isentropic efficiency is observed in the optimized geometry. However, it should be noted that efficiency maximization is not picked as an objective. This situation is thought to be in correlation with the increment in the tip camber observed in the optimum geometry. Besides, the overall increase in stall margin, optimized geometry's enhanced pressurizing capability, and more comprehensive working range show that genetic algorithm provided an optimized design while respecting constraints.

Conclusions

In this study, optimization of a baseline axial compressor inspired from NASA Rotor 37 is performed. An in-house optimization code, which utilizes a genetic algorithm is coupled with the throughflow design software AxStream. Aerodynamically optimum geometry is obtained. A remarkable improvement in the objective function is observed, which in turn provided a compressor blade that is capable of working on a more comprehensive operation margin for design RPM with a higher pressurization capability.

Acknowledgments

The authors would like to thank TUSAŞ Engine Industries (TEI) for their support.

Nomenclature

DF	Diffusion Factor
$\bar{f}$	Normalized objective function
$f_i(\bar{x})$	Objective function vector
F_q	Local blade force along quasi-orthogonal
$g_j(\bar{x})$	Constraint function vector

$\dot{m}$	Mass flow rate passing through the compressor
P_0	Stagnation pressure
$PR_{t\text{-}t}$	Blade exit to inlet pressure ratio based on total pressures
q	Local direction along quasi-orthogonal
r	Radius
s	Entropy
T	Temperature
r_k	Penalty multiplier
w_1	Relative velocity at the leading edge
w_2	Relative velocity at the trailing edge
$\bar{x}$	Design variable vector
V_θ	Tangential velocity
ε	Local angle between streamline true and quasi-orthogonal
η_{is}	Isentropic efficiency
ϕ	Pseudo objective function
σ	Solidity
H	Total enthalpy
K_m	Local streamline curvature
SM	Stall margin
V_m	Meridional velocity

References

[1] Ellbrant, L. Eriksson, L. and Martensson, H. 2012. Design of Compressor Blades Considering Efficiency and Stability Using CFD Based Optimization. ASME Turbo Expo 2012, Copenhagen, Denmark, ASME.

[2] Ji, L. Yi, W., Li, W., Tian, Y. and Chen, J. 2012. Multi-Stage Turbomachinery Blades Optimization Design Using Adjoint Method and Thin Shear-Layer N-S Equations. ASME Turbo Expo 2012, Copenhagen, Denmark, ASME.

[3] Dawes, W. 2007. Turbomachinery computational fluid dynamics: asymptotes and paradigm shifts. Phil. Trans. R. Soc. A. 365: 2553–2585.

[4] Kor, O. 2016. Aerodynamic Optimization of a Transonic Aero-Engine Fan Module. Ph.D. Thesis, İzmir Institute of Technology, İzmir, Turkey.

[5] Oyama, A., Liou, M. and Obayashi, S. 2002. Transonic Axial-flow Blade Shape Optimization Using Evolutionary Algorithm and Three-Dimensional Navier Stokes Solver. 9th AIAA/ISSMO Symposium on Multidisciplinary Analysis and Optimization, Atlanta, Georgia, USA, AIAA/ISSMO.

[6] Giannakoglou, K. and Papadimitriou, D. 2008. Adjoint methods for shape optimization. pp. 79–108. *In:* Thévenin, D. and Janiga, G. [eds.]. Optimization and Computational Fluid Dynamics, Springer, Heidelberg, Germany.

[7] Hannah, L. Stochastic Optimization. 4 April 2014. [Online]. Available: http://www.stat.columbia.edu/~liam/teaching/compstat-spr14/lauren-notes.pdf. [Accessed 27 July 2019].

[8]   Lian, Y. and Liou, M. 2008. Multi-objective optimization of transonic compressor blade using evolutionary algorithm. Journal of Propulsion and Power 21: 979–987.

[9]   Verstraete, T., Müller, L. and Aissa, M. 2018. Multidisciplinary optimization of turbomachinery components using differential evolution. pp. 3–76. *In*: Périaux, J. and Verstraete, T. [eds.]. Introduction to Optimization and Multidisciplinary Design. VKI LS, Sint-Genesius-Rode, Belgium.

[10] Chatel, A., Verstraete, T. and Coussement, G. 2019. Multipoint Optimization of an Axial Turbine Cascade Using A Hybrid Algorithm. ASME Turbo Expo 2019, Phoenix, Arizona, USA, ASME.

[11] Sudeepta, M., Joly, M. and Sarkar, S. 2019. Multi-Fidelity Global-Local Optimization of a Transonic Compressor Rotor. ASME Turbo Expo 2019, Phoenix, Arizona, USA, ASME.

[12] Mengitsu, T. and Ghaly, W. 2008. Aerodynamic optimization of turbomachinery blades using evolutionary methods and ANN-based surrogate models. Optimization and Engineering. 9: 239–255.

[13] Deb, K. 2010. Optimization for Engineering Design—Algorithms and Examples. PHI Learning Private Ltd., New Delhi.

[14] Rao, S. 1996. Engineering Optimization, Theory and Practice [3rd ed.]. John Wiley and Sons Inc., New York.

[15] Resmini, M. 2012. Numerical Investigation of the Solidity Effect on a Linear Compressor Cascade Performance and Stability. M.S. Thesis. Politecnico di Milano, Milano, Italy.

[16] Davidson, P. 2015. Turbulence [2nd ed.]. Oxford University Press, Oxford, United Kingdom.

[17] ANSYS CFX-Solver Theory Guide. 2017. ANSYS Inc. Software Documentations for Release 18.2. Pennsylvania.

[18] Menter, F. 2009. Review of the shear-stress transport turbulence model experience from an industrial perspective. International Journal of Computational Fluid Dynamics 23: 305–316.

[19] Suder, K. 1996. Experimental Investigation of the Flow Field in a Transonic, Axial Flow Compressor with Respect to the Development of Blockage and Loss. NASA TM 107310, Cleveland, Ohio.

[20] Shahpar, S. 2012. Optimization strategies used in turbomachinery design from an industrial perspective. pp. 1–22. *In*: Périaux, J. and Verstraete, T. [eds.]. Introduction to Optimization and Multidisciplinary Design in Aeronautics And Turbomachinery. VKI LS, Sint-Genesius-Rode, Belgium.

[21] Joly, M., Verstraete, T. and Paniagua, G. 2012. Full Design of a Highly Loaded Fan By Multi-Objective Optimization of Through-Flow and High-Fidelity Aero-Mechanical Performances. ASME Turbo Expo 2012, Copenhagen, Denmark, ASME.

[22] Cumpsty, N. 2004. Compressor Aerodynamics. Krieger Publishing Company, Florida.

[23] Acarer, S. and Özkol, Ü. 2015. Development of A New Universal Inverse Through Flow Program and Method for Fully Coupled Split Flow Turbomachinery Systems. ASME Turbo Expo 2015, Montréal, Canada, ASME.

[24] Giri, G., Nassar, A., Moroz, L., Klimov I. and Sherbina, A. 2016. Design and Analysis of a High Pressure Ratio Mixed Flow Compressor Stage. 52nd AIAA/SAE/ASEE Joint Propulsion Conference, Burlington, U.S.A., AIAA/SAE/ASEE.

[25] Lieblein, S. 1965. Experimental flow in 2D cascades. pp. 1–99. *In*: Bullock, R.O. [ed.]. Aerodynamic Design of Axial Flow Compressors. NASA SP-36. Washington DC, USA.

CHAPTER 15

Heat Transfer Optimization by Multiple Nonlinear Neuro-regression Approach for Ground Source Heat Exchanger

Şahin Güngör[1],* and *Murat Aydin*[2]

Introduction and Literature Survey

In parallel with the development of computer technology and industrial manufacturing techniques, designs are becoming more original and more aesthetic, and functional outputs are getting widespread. Nowadays, the human population and their energy consumption is getting incredibly high. Therefore, we continuously hear the words for more efficient, lighter, cheaper, or longer-lasting. It is essential to understand that the most efficient demands are through the optimization work that is put forward with the right techniques. On the other hand, due to the rapid increase in energy consumption in parallel with the increase of population, both renewable energy sources and energy conservation are trending topics in the world. Energy-saving, which has become a state policy in developed countries, is also essential to obtain energy independence and to protect nature [1]. Academia and industrial R&D centers are constantly looking for energy-efficient technologies to provide energy production with renewable energy sources to minimize energy consumption.

[1] İzmir Kâtip Çelebi University, Department of Mechanical Engineering, İzmir, Turkey.
[2] EMG Mechanics, Electric, Natural Gas, Construction and Assembly Inc., Aydın, Turkey.
   Email: emgmekanik@gmail.com
* Corresponding author: sahin.gungor@ikc.edu.tr

Heat exchangers are the most widely used components in thermal-fluid engineering systems, especially for energy recovery, and the reuse of waste energy. They are used in many applications, ranging from combi-boilers, air-conditioners, and the refrigerators in our homes to petroleum and natural gas plants in industrial facilities. Heating and cooling systems are inevitable in terms of thermal comfort and indoor air quality. Although there are lots of alternative systems for cooling and heating, refrigeration and heat pump cycles are the most widespread thermodynamic solutions in HVAC (heating, ventilation and air conditioning) technologies. Heat pump systems are preferred due to their higher thermal performance. Their interaction with air, water, and soil resources depending on the energy storage determine their performance. Geothermal energy, which is one renewable energy source, is widely used in regions with high a geothermal potential. However, the idea of using underground water as a heat source is spreading recently.

Additionally, soil and groundwater sources are especially important for heat pump systems for heat in cold days. These heat pumps work best in regions where summer and winter average temperature values are very close to each other. Soil or groundwater heat pump systems can be applied horizontally or vertically. Their heat transfer capabilities can determine the thermal efficiency of underground heat exchanger systems used as heat pumps.

Several models have been reported for the heat transfer in ground heat exchangers and heat pump systems. Most of the studies are based on either analytical or numerical methods to analyze the thermal-fluid engineering problems. [2, 3] used a finite line source model. They developed an analytical expression by calculating the ground temperature variation. Besides, they examined the thermal performance of the borehole systems using vertical ground heat exchangers. 3D models were used to simulate ground heat exchangers [4, 5, 6] to estimate their thermal resistance and best-layered ground profiles.

Previous studies [6, 7] examined the heat transfer distribution and transient temperature variation. Starace [8], used experimental and numerical methods to evaluate a horizontal ground source heat pump's thermal performance. The main parameters affecting the thermal performance were investigated in another study [9]. Fluent, a commercial computational fluid dynamics (CFD) software, was used for simulating the heat pump's characteristics. Yang et al. [10], used both the steady and transient heat-transfer method to analyze the heat transfer characteristics of a model of a vertical U-tube ground type heat exchanger. They validated their results with an experiment. Li and Zheng [11], also developed a three-dimensional finite volume domain for modeling of a vertical ground source heat exchanger. Borehole domain was divided into many layers to obtain the thermal effects in the vertical direction accurately. The results of the vertical borehole model were compared with experimental data to examine the accuracy of the 3D model. Zhang et al. [12], presented a new design model by considering the uncertainty in the borehole thermal resistance and thermal conductivity of the heat pump. The thermal response test method was used to determine thermal parameters. Critical issues in parameter estimation during the thermal response test process were also discussed. Poulopatis et al. [13], determined the factors affecting the sizing and positioning of Ground Heat Exchangers (GHEs) in Cyprus. The influences of the temperature, thermal conductivity, specific heat and density of the ground as well as pipe diameter on the performance of GHEs were investigated by using a numerical model in conjunction with the test data. Ahsaee and Ameri [14], used a numerical model that was extended to analyze the energy and exergy of direct-expansion geothermal heat pumps. The results of this study are used to design and optimize a direct-expansion geothermal heat pump.

It is possible to create more efficient systems by applying optimization, which is one of the more recent topics nowadays with regards to heat exchanger and heat pump systems. Esen et al. [15–20], studied in detail about the performance optimization of ground-coupled heat pump systems. They performed an optimization study with different methods such as neural network, and neuro-fuzzy inference systems. Sanaye and Modarrespoor [21], thermally modeled the heat exchanger system by using a ε-NTU method and used a genetic algorithm to optimize the overall heat transfer coefficient of the heat exchanger. Besides that, they also took into account cost parameters. Multi-objective improved teaching-learning-based optimization (MO-ITLBO) algorithm was performed for a plate-fin heat exchanger model by Patel and Savsani [22]. They aimed to minimize the pressure drop while maximizing the plate-fin type heat exchanger effectiveness. Yang et al. [23], also used a general optimization design method motivated by constructal theory. This method was proposed for the heat exchanger design and optimization process.

Heat sinks are the heat exchange systems for cooling the electronic components by the help of different kinds of fins. Ahmed et al. [24] summarized the heat sink systems in a review paper. Heat sinks are evaluated in terms of their thermal performance and limitations. Raja et al. [25], investigated the thermal-hydraulic performance of a plate heat exchanger by multi-objective optimization. Multi-objective optimization was aimed to maximize the overall heat transfer coefficient and minimize the pressure drop of the plate type heat exchanger. The optimization model and the experimental results were very close to each other, and the deviations were approximately 9%.

In another study [26], a 3-dimensional, unsteady, and numerical model of a vertical ground heat exchanger was modeled by using FLUENT computational fluid dynamics (CFD) software. There was data set for nine independent variables for the heat exchanger which was optimized. By using a nonlinear multiple regression model, heat transfer per unit length was calculated for different values and independent variables.

Method

Multiple Nonlinear Regression Model

Regression analysis is used for estimating the relationships between the parameters of an engineering problem [27]. The variables that are desired to predict outcomes by using a linear or nonlinear regression model are called target or dependent variables. The independent variables, which can be considered as design variables, are used to predict the dependent variables. We can fit the data set by different multiple regression models, and the impact of each design variable (predictors) to the total phenomena can also be determined. There are different kinds of multiple regression models such as linear, nonlinear, rational, and logistic. Linear regression models are simple in terms of defining and solving engineering problems. However, real working patterns include nonlinear behaviors (such as logarithmic, power, trigonometric, and rational forms). Rational regression models have many advantages. They have better interpolator properties than polynomial models, and they have excellent asymptotic properties. They can also be used to model complicated structures [27, 28]. The independent predictors related to the dependent variable, can be expressed in the correlation coefficient '*R*,' which is the square root of '*R-square*' value. In multiple regressions, this model can be examined by statistical methods and stability tests. The details about nonlinear regression models and the algorithms are presented in previous chapters of the book to achieve intelligibility and clarity.

Multiple Nonlinear Neuro-regression Approach

In the optimization of a thermal-fluid engineering problem, an artificial neural network (ANN) approach is coupled with multiple nonlinear regression models to check the accuracy of the predictions. In this approach, 20% of data is considered as testing data, and it is randomly extracted from the whole data set. 80% of the remaining data is used for training purposes, which aims to minimize the error between the exact points and the model results. In the Multiple Nonlinear Neuro-regression approach, the objective function of the optimization problem is created by training the data. Then, the model is checked with the testing data in terms of boundedness criteria. Checking the boundedness of regression or hybrid models helps us to know whether the model satisfies the phenomena of the data set or not. Samples of the boundedness check is shared in a further section, in detail.

Results and Discussion

In the reference work [26], a vertical ground heat exchanger is modeled, and the results of the finite element analysis were examined statistically. The heat exchanger was optimized using a multiple regression model. In their model, heat transfer per unit length (dependent variable) was defined by nine independent (design) variables, as shown in Table 15.1.

Table 15.1 Independent variables used for optimizing the heat pump system.

Independent Variables	Physical Meaning	Unit
λ_{soil}	Soil thermal conductivity	W/(m.K)
$(\rho Cp)_s$	Soil volumetric heat capacity	kJ/$(m^3.K)$
λ_{bulk}	Fluid thermal conductivity	W/(m.K)
$(\rho Cp)_b$	Fluid volumetric heat capacity	kJ/$(m^3.K)$
V_{in}	Inlet flow rate	m^3/h
T_{in}	Inlet Temperature	K
φ	Porosity	–
u	Groundwater flow	m/day
H	Borehole depth	m

Evaluation of Numerical Data Set

In optimization analysis, the data set obtained by experimental, numerical, or theoretical methods is needed. In the reference study, the results were obtained for the vertical ground heat exchanger and analyzed by the computational fluid dynamics. As a result of the numerical analysis, the heat transfer amount per unit length was calculated for 64 cases in the reference study [26]. Table 15.2 shows a set of numerical data samples used in regression and optimization analyses.

Determination of the Multiple Nonlinear Neuro-regression Model

The correlations between heat flux and each parameter that passed the monadic regression test are used in optimization analysis. It is found that heat flux depends on all other parameters except soil porosity and borehole depth. Linear relationships are observed between the backfill volumetric

Table 15.2 Experimental data set sample [26].

No.	λ_{soil} W/(mK)	$(\rho Cp)_s$ kJ/(m³.K)	λ_{bulk} W/(mK)	$(\rho Cp)_b$ kJ/(m³.K)	V_{in} m/h	T_{in} K	φ	U m/day	H m	q_L W/m
1	2.9	3780	2	4950	0.246	298	0.39	172.8	50	23.27
2	4.21	1620	2.8	2400	0.246	308	0.24	0.0172	70	45.44
3	2.9	4320	2.8	4050	0.308	298	0.24	0	70	22.93
4	2.4	1620	2.8	2400	1.231	298	0.39	1.728	50	57.52
5	3.55	2430	2.4	1725	0.246	308	0.6	0	50	64.10
6	2.4	1620	2.4	4950	0.308	313	0.6	1.728	30	122.68
7	4.21	2430	1.6	1725	0.615	298	0.51	1.728	50	41.42
8	1.75	1620	1.6	1725	0.246	298	0.12	0	30	24.62
9	2.4	4320	1.6	3075	0.923	308	0.39	0.0172	30	82.11
10	2.9	2970	2.8	1725	0.308	303	0.12	172.8	30	102.86
11	3.55	2970	1.6	4950	1.231	303	0.24	0	70	65.38
12	1.75	1620	1.6	1725	0.246	298	0.12	0	30	24.64
13	2.4	3780	2.4	3075	1.231	298	0.12	0.0172	70	40.45
14	1.75	2430	2	4050	1.231	308	0.51	1.728	30	112.68
15	2.4	4320	1.6	17250	0.615	298	0.24	1.728	100	30.58
16	2.4	3780	2.4	1.725	0.246	308	0.24	0	120	26.36

17	1.75	2970	2.4	3075	0.923	298	0.24	172.8	50	77.74
18	1.75	2430	1.6	2400	0.615	313	0.24	172.8	70	135.57
19	2.9	2430	2.4	2400	0.246	318	0.24	17.28	30	173.82
20	2.9	2970	2.4	1725	0.615	308	0.39	1.728	70	95.98
21	3.55	1620	2.4	4050	0.615	318	0.12	0.0172	50	147.05
22	2.9	3780	1.6	2400	0.308	318	0.12	1.728	50	115.54
23	1.75	1620	2	1725	0.923	318	0.39	0.0172	70	101.7
24	1.75	2970	2	3075	0.246	318	0.24	1.728	100	54.57
25	1.75	3780	3.2	4050	0.308	308	0.24	1.728	30	102.42

heat capacity, inlet water temperature, soil porosity, and borehole depth. On the other hand, exponential relationships are observed between the soil and backfill thermal conductivity, soil volumetric heat capacity, inlet flow, and groundwater flow velocity. Due to the high R square values at the end of the correlation, nine design variables are used to predict the borehole heat flux. The decision of the regression model depends on the monadic regression approach by in-house software to see the trend of each design variables in good agreement with the reference study. The general form of the regression model is given in Equation (15.1)

$$q_l = f(a * b^{\lambda_{soil}} + c^{(\rho Cp)_s} + d * e^{\lambda_{bulk}} + f * (\rho Cp)_b + g * h^{V_{in}} + i * T_{in}$$
$$+ j * \varphi + k * l^u + m * H) \tag{15.1}$$

where, a-m are the coefficients of regression analysis and the λ_{soil}, $(\rho Cp)_s$, λ_{bulk}, $(\rho Cp)_b$, V_{in}, T_{in}, φ, u and H are the independent design variables, and q_l is the dependent variable of the optimization analysis. Similar to the reference study, the effect of soil porosity and volumetric heat capacity on heat transfer is ignored. After the regression analysis, the objective function of the vertical ground heat exchanger can be expressed as:

$$q_l = -1236.359 - 94.006 * 0.238^{V_{in}} - 113.563 * 0.292^{\lambda_{bulk}} -$$
$$58.918 * 0.772^{\lambda_{soil}} - 42.949 * 0.934^u + 4.740 * T_{in} - 0.386 * H \tag{15.2}$$

Table 15.3 The statistical summary of the regression model for all data sets.

R^2	R^2_{adj}
0.8727	0.8593

The Artificial Neural Network (ANN) approach is also examined with the multiple stepwise regression model. For this hybrid approach, the numerical data is divided into two random parts, firstly nearly 80% (54 lines) and 20% (10 lines) are randomly selected, and they are used as testing data in the model that is created with data set of 54 lines. The objective function of the vertical ground heat exchanger for a data set with 54 lines can be expressed as:

$$q_l = -1056.492 - 95.720 * 0.238^{V_{in}} - 163.457 * 0.292^{\lambda_{bulk}} - 88.856 * 0.772^{\lambda_{soil}} -$$
$$40.557 * 0.934^u + 4.219 * Tin - 0.431 * H \tag{15.3}$$

Table 15.4 The statistical summary of the regression model for the training data set.

R^2	R^2_{adj}
0.8034	0.8533

The ANN hybrid approach is applied to the regression model by randomly selected testing data. Testing data is applied in the model generated from the data set with 54 design point. Table 15.5 shows the statistical summary of the hybrid regression model.

In Table 15.5, R-square and adjusted R square values are over for 80% of both the whole data set and testing data set created by ANN hybrid method. In particular, high R square value is calculated by using the test data, which indicates that the model well describes the data points.

Table 15.5 The statistical summary of the regression model for testing the data set.

R^2	R^2_{adj}
0.8430	0.7645

Boundedness Check for the Regression Model

Boundedness check is a necessary way to see how well the model describes the phenomena. For a well-described model, the R-square value should converge to 1. For boundedness control, random values must be selected between any two consecutive points in the data set. For example, for "case 1", the independent variable values are selected (given in Table 15.6). All the selected values are represented in Table 15.7 and can be calculated easily with the objective function provided in Equation (15.3).

The boundedness test results show that the ANN hybrid regression model is successful in covering the data set and satisfying the stability of objective function.

Table 15.6 Boundedness control test sample.

Lines	25th	Between 25th–26th	26th
q_l	112.68	68.2486	30.58

Thermal Optimization of Ground Heat Exchanger

Determination of Design Variables and Constrains

As the design variables of the vertical ground heat exchanger, nine independent variables are determined as provided in Table 15.2. These variables must be optimized to maximize the amount of heat transfer per unit length. In this way, it is possible to design and use the system in the most efficient way. However, it must be checked whether the design variables are in discrete form, and the constraints are defined during the analysis. Table 15.8 shows the constraints for the design variables which are determined according to the physics of the problem.

Table 15.7 Boundedness check for the multiple regression model (All cases include random values).

Parameters	Case 1	Case 2	Case 3	Case 4	Case 5	Case 6	Case 7	Case 8	Case 9	Case 10
k_{soil} (W/m.K)	1.90	2.525	2.6178	3.135	3.77	2.71	3.3333	2.981	2.655	3.193
$(\varrho C_p)_{soil}$	3200	2430	3123.42	1999.91	2944.44	2970	2430	1620	1620	1620
k_{bulk} (W/m.K)	1.85	2.998	1.7222	2.8	1.712	1.6	2.4	2.833	1.982	2.13
$(\varrho C_p)_{bulk}$	2625	3888	4312.02	2716	2233	4950	2711	3227.9	3777.78	4001.21
V_{in} (m³/h)	0.813	0.4357	0.246	0.246	0.3871	0.8888	0.297	0.554	0.888	0.615
T_{in} (K)	303	311	300.98	305.13	301.05	307.98	306.66	303	304.44	315.91
ϕ	0.37121	0.39	0.39	0.418	0.3612	0.4109	0.1777	0.3111	0.163	0.182
u_{water}	1.728	0.01	21.666	0.813	1.728	0	17.28	9.56	54.54	0.01
H (m)	43.0413	72.098	117.93	66.66	59.33	70	45.59	53.77	71.1234	99.87
q_l (W/m)	68.2486	83.3397	25.4322	47.8582	39.7024	82.2797	94.3650	83.6960	112.4710	106.7330
Tolerance	Within Range	Within Range	9% bias	5% bias	Within Range	8% bias	Within Range	Within Range	Within Range	Within Range

Table 15.8 Constraints of the systems.

$\lambda_{soil} \in \{1.75, 2.4, 2.9, 3.55, 4.21\}$
$(\rho Cp)_s \in \{1620, 2430, 2970, 3780, 4320\}$
$\lambda_{bulk} \in \{1.6, 2, 2.4, 2.8, 3.2\}$
$(\rho Cp)_{bulk} \in \{1725, 2400, 3075, 4050, 4950\}$
$V_{in} \in \{0.246, 0.308, 0.615, 0.923, 1.231\}$
$T_{in} \in \{298, 303, 308, 313, 318\}$
$\varphi \in \{0.12, 0.24, 0.39, 0.51, 0.6\}$
$u \in \{0, 0.01728, 1.728, 17.28, 172.8\}$
$H \in \{30, 50, 70, 100, 120\}$

Optimization Study Results

Optimization is the basis of minimization or maximization of the objective function, which is obtained by the multiple regression model and is understood to adequately describe the phenomena as a result of the necessary controls. The objective function (equation two) is obtained by using the multiple regression and by taking into account the constraints provided in Table 15.8. "Nelder-Mead", "Random Search", "Differential Evolution" and "Simulated Annealing" algorithms in the Mathematica program are used to optimize the ground source heat exchanger. Table 15.9 shows the optimized values for the system parameters.

Table 15.9 shows the optimal values for the soil and backfill thermal conductivity, inlet flow rate, inlet water temperature, groundwater flow velocity, borehole depth, and heat transfer per unit length. Optimization study indicates that heat transfer at a rate of approximately 220 W per

Table 15.9 Optimized parameters for the vertical ground source heat pump.

λ_{soil}	λ_{bulk}	V_{in}	T_{in}	u	H	q_l
4.21	3.2	1.231	318	172.8	30	221.426

Optimization Method	q_l Value (W/m)	Simulation Time (s)
Nelder-Mead Algorithm (Deterministic)	221.426	726.26
Differential Evolution Algorithm (Stochastic)	221.426	742.89
Simulated Annealing Algorithm (Stochastic)	221.426	770.57
Random Search Algorithm (Stochastic)	221.426	751.21

unit length can be achieved for with this vertical heat pump problem. The variables that this has have a minimal effect and are not included for the hybrid regression model. Therefore, porosity and volumetric heat capacity are neglected during the optimization process. The optimization result shows that the output (heat flux per unit length) has reached its maximum value when:

- *thermal conductivity of soil is maximum,*
- *thermal conductivity of bulk is maximum,*
- *inlet velocity is maximum,*
- *inlet water temperature is maximum,*
- *underground volumetric flow is maximum,*
- *borehole depth is minimum.*

As shown in Table 15.9, the results obtained with the stochastic or deterministic optimization algorithms are the same. The results are the same because we have obtained a simple mathematical expression (objective function) by ANN hybrid multiple regression analysis. It is also observed that the fastest solution is obtained based on the Nelder-Mead algorithm, while the slowest one is obtained based on the Simulated Annealing algorithm.

Conclusion

A 3D numerical model for a vertical ground heat exchanger is simulated in the reference study using nine different input parameters. The effects of thermal conductivity, volumetric heat capacity, inlet flow, inlet water temperature, soil porosity, underground water flow rate, and bore-hole depth on heat flux are investigated. In this study, the design variables of the heat pump are optimized; so that the system can operate most efficiently. How well the multiple regression model describes the phenomena depends on the reliability of optimization data set.

The Multiple Nonlinear Neuro-regression approach is used, and the regression model outputs are tested for the points that are not included in the data set by boundedness tests. Additionally, how well the generated hybrid regression model describes the points in the data set and phenomena of the thermal-fluid engineering problem is investigated. Stochastic optimization algorithms of Differential Evolution, Random Search, Simulated Annealing, and Nelder-Mead are used as optimization methods in the Mathematica software. For each algorithm, the amount of heat transfer per unit length and the duration of the solution are calculated. It is observed that the optimized condition is obtained for a situation which is not included in the data set.

References

[1] Karagöl, E.T. and Kavaz, I. 2017. Dünyada ve Türkiye'de yenilenebilir enerji, SETA. 197.

[2] Zeng, H.Y., Diao, N.R. and Fang, Z.F. 2002. A finite-line source model for boreholes in geothermal heat exchangers. Heat Tran Asian Res. 31: 558–567.

[3] Zeng, H.Y., Diao, N.R. and Fang, Z.F. 2003. Heat transfer analysis of boreholes in vertical ground heat exchangers. Int. J. Heat. Mass Transf. 46: 4467–4481.

[4] Yoon, S., Lee, S.R. and Go, G.H. 2014. A numerical and experimental approach to the estimation of borehole thermal resistance in ground heat exchangers. Energy 71: 547–555.

[5] Luo, J., Rohn, J., Bayer, M., Priess, A. and Xiang, W. 2014. Analysis on performance of borehole heat exchangers in layered subsurface. Appl. Energy 123: 55–65.

[6] Raymond, J., Therriena, R. and Gosselin, L. 2011. Borehole temperature evolution during thermal response tests. Geothermics 40: 69–78.

[7] Özüdoğru, T.Y., Olgun, C.G. and Senol, A. 2014. 3D numerical modelling of vertical geothermal heat exchangers. Geothermics 51: 312–324.

[8] Starace, G., Congedo, P.M. and Colangelo, G. 2005. Horizontal heat exchangers for GSHPs; Efficiency and cost investigation for three different applications. The Conference: ECOS2005-18th International Conference on Efficiency, Cost, Optimization, Simulation and Enviroment, Trondheim.

[9] Starace, G., Congedo, P.M. and Colangelo, C. 2012. CFD simulations of horizontal ground heat exchangers: a comparison among different configurations. Appl. Therm. Eng. 33: 24–32.

[10] Yang, W.B., Shi, M.H. Liu, G.Y. and Chen, Z.Q. 2009. A two-region simulation model of vertical U-tube ground heat exchanger and its experimental verification. Appl. Energy. 86: 2005–2012.

[11] Li, Z.J. and and Zheng, M.Y. 2009. Development of a numerical model for the simulation of vertical U-tube ground heat exchangers. Appl. Therm. Eng. 29: 920–924.

[12] Zhang, X., Huang, G., Jiang, Y. and Zhang, T. 2015. Ground heat exchanger design subject to uncertainties arising from thermal response test parameter estimation. Energy Build. 102: 442–452.

[13] Pouloupatis, P.D., Tassau, S.A. Christodoulides, P. and Florides, G.A. 2017. Parametric analysis of the factors affecting the efficiency ground heat exchangers and design application aspect in Cyprus. Renew Energy. 103: 721–728.

[14] Ahsaee, H.G. and Ameri, M. 2018. Energy and exergy investigation of a carbon dioxide direct-expansion geothermal heat pump. Appl. Therm. Eng. 129: 165–178.

[15] Esen, H., Inalli, M., Sengur, A. and Esen, M. 2008a. Modelling a ground-coupled heat pump system using adaptive neuro-fuzzy inference systems. Int. J. Refrig. 31: 65–74.

[16] Esen, H., Inalli, M., Sengur, A. and Esen, M. 2008b. Artificial neural networks and adaptive neuro-fuzzy assessments for ground-coupled heat pump system. Energy Build. 40: 1074–1083.

[17] Esen, H., Inalli, M., Sengur, A. and Esen, M. 2008c. Forecasting of a ground-coupled heat pump performance using neural networks with statistical data weighting pre-processing. Int. J. Therm. Sci. 47: 431–441.

[18] Esen, H., Inalli, M., Sengur, A. and Esen, M. 2008d. Modelling a ground-coupled heat pump system by a support vector machine. Renew Energy. 33: 1814–1823.

[19] Esen, H., Inalli, M., Sengur, A. and Esen, M. 2008e. Performance prediction of a ground-coupled heat pump system using artificial neural networks. Expert Syst. Appl. 35: 1940–1948.

[20]   Esen, H., Inalli, M., Sengur, A. and Esen, M. 2008f. Predicting performance of a ground-source heat pump system using fuzzy weighted pre-processing-based ANFIS. Build Environ. 43: 2178–2187.

[21]   Sanaye, S. and Modarrespoor, D. 2014. Thermal-economic multi-objective optimization of heat pipe heat exchanger for energy recovery in HVAC application using genetic algorithm. Therm. Sci. 18: 375–391.

[22]   Patel, V. and Savsani, V. 2014. Optimization of a plate-fin heat exchanger design through an improved multi-objective teaching-learning based optimization (MO-ITLBO) algorithm. Chem. Eng. Res. Des. 92: 2371–2382.

[23]   Yang, J., Oh, S.R. and Liu, W. 2014. Optimization of a shell and tube heat exchanger using a general design approach motivated by constructal theory. Int. J. Heat. Mass. Transf. 77: 1144–1154.

[24]   Ahmed, H.E., Salman, B.H., Kherbeet, A.S. and Ahmed, M.I. 2018. Optimization of thermal design of heat sinks: A review. Int. J. Heat. Mass. Transf. 118: 129–153.

[25]   Raja, B.D., Jhala, R.L. and Patel, V. 2018. Thermal-hydraulic optimization of plate heat exchanger: A multi-objective approach. Int. J. Therm. Sci. 124: 522–535.

[26]   Chen, S., Mao, J. and Han, X. 2016. Heat transfer analysis of a vertical ground heat exchanger using numerical simulation and multiple regression model. Energy Build. 129: 81–91.

[27]   Özturk, S., Aydın, L., Küçükdoğan, N. and Çelik, E. 2017. Optimization of lapping process of silicon wafer for photovoltaic application. Sol. Energy 164: 1–11.

[28]   Rao, S.S. 2009. Engineering Optimization: Theory and Practice. John Wiley & Sons, London.

CHAPTER 16

Thermal Optimization of Lightweight and Micro-porous Clay Bricks for Building Applications

Savas Ozturk,[1] *Mucahit Sutcu,*[2,*]
Levent Aydin[3] *and Osman Gencel*[4]

Introduction

In order to improve the thermal performance of clay-based brick materials, which is the oldest known building material used in buildings, it is aimed to form a microporous structure in the brick bodies with the help of the additives used in the production process and thus reduce the thermal conductivity values of the brick material. Today, many organic and inorganic pore-making additives are blended with clay raw materials to produce porous clay bricks that give to brick and both lighten its weight and lower thermal conductivity. In terms of conservation of energy the thermal insulation of buildings which will contribute to the

---

[1] Department of Metallurgical and Materials Engineering, Manisa Celal Bayar University, Manisa, 45140, Turkey.
  Email: savas.ozturk@cbu.edu.tr
[2] Department of Materials Science and Engineering, Izmir Katip Celebi University, Izmir, 35620, Turkey.
[3] Department of Mechanical Engineering, Izmir Katip Celebi University, Izmir, 35620, Turkey.
  Email: leventaydinn@gmail.com
[4] Department of Civil Engineering, Bartin University, Bartin, 74100, Turkey.
  Email: osmangencel@gmail.com
* Corresponding author: mucahit.sutcu@ikc.edu.tr

low thermal conductivity values are very important. In this chapter, the thermal optimization of porous and lightweight clay bricks produced by using agricultural waste is studied. In this study, the Box-Behnken experimental design was chosen in determining the experimental set, and response surface analyses were performed. Milled nut pine shells (Pinus pinea) in powder form are used in the production of bricks as the pore maker organic additive and added to clay raw material at two different ratios of 5% and 10% by weight. Green brick samples were produced from mixtures prepared in different ratios; the powder mixtures were compressed in the mold by the hydraulic press and fired at three different firing temperatures and times (850, 950, 1050°C and 2, 4, 6 hours) after the drying step. The physical, mechanical, and thermal properties of the fired brick samples were determined. Amount of additives, firing times, and firing temperatures data were used to create a nonlinear mathematical model based on the Neuro-regression approach. Then thermal conductivity values were correlated with firing parameters, and the predictability of firing parameters was investigated. Finally, DE, NM, SA, and RS algorithms were used in the optimization of the thermal conductivity of bricks in the firing process.

Literature Survey

Clay-based bricks are widely used as construction materials worldwide, with an annual production of approximately 1391 billion units [1]. These materials are preferred mostly because of their lower cost and high fire resistance, high durability, and longer service life [2, 3]. In industrial processing, the manufacturing of clay bricks is made up of several stages such as extraction of the raw materials, preparation of the clay body, shaping of the products, drying, firing, and treatments after firing. The clay-based brick and tile products are typically used in the constructional applications. These products are generally solid bricks that have a rectangular shape, bricks with vertical perforations and roof tiles depending on the application purposes. The voids of perforations in the bricks reduce the quantities of clay raw materials required and increase the thermal insulation performance of the bricks [3]. Meanwhile, the high demand for clay raw materials in brick production has caused excessive use and thus decline of this non-renewable raw material in some of the world's fastest-growing countries [4, 5]. Therefore, it is essential to evaluate the environmentally friendly alternative additives such as organic and inorganic wastes in order to replace the clay raw materials in the brick production due to the decrease of clean clay deposits.

Nowadays, increasing global industrial activities cause critical environmental problems. Specifically, a significant amount of waste material is generated from different industrial sources. Many of these wastes are generally dumped in landfills for their disposal. However, an increasing amount of waste material requires large storage areas that are difficult to maintain. Therefore, the re-use or recycling of industrial and agricultural wastes is essential not only as an economic opportunity worldwide but also for solving environmental problems. Most of these waste materials can be reused in the building materials industry. For example, since clay-based fired bricks consist of raw materials with a wide range of compositions, they can tolerate composition changes due to the presence of different types of waste even at high rates [6, 7].

There has been considerable research on the use of different types of wastes for making various constructional materials. Silicate based wastes such as coal and fly ash, slag from steel production, hydrometallurgical mud, and different types of industrial sludge are the proper materials for using in the ceramic and cement industries due to their chemical and mineralogical contents. Recently, the utilization and applications of pore making wastes have gained importance due to their organic and inorganic content, and many studies are still ongoing for further uses [8, 9, 10, 11]. A study on the optimization of inorganic mineral wastes content in traditional ceramics using a statistical design of mixture experiments was employed [12]. Regression models were used to optimize the waste content in ceramic body compositions, and it was concluded that formulations containing up to 62% waste could be used.

In this chapter, a thermal optimization study of porous and lightweight clay bricks produced by using an agricultural residue is performed. It is a very suitable material for the production of clay-based porous bricks. This residue provides excellent physical and thermal properties such as high porosity, low density, and low thermal conductivity to the products. The bricks produced with the incorporation of this residue can be utilized in building applications for thermal insulation.

Objectives and Motivation of the Study

Microporous clay-based bricks must have a particular mechanical strength and a density that does not increase the weight of the total mass in order to be used as a building material. In addition to these basic properties, the thermal conductivity of the brick should be low, i.e., it is desirable to exhibit good heat-insulating behavior. This insulating property can be achieved by the hollow shaping of bricks, as well as by adding organic or inorganic additives to the clay blend that burns at a high temperature and leaves a gap in the area where it is located.

These admixtures must have a homogeneous distribution in the brick mix in ratio and size to keep the mechanical strength within practical limits. While various inorganic based industrial process wastes are used as a pore former, many kinds of research have been conducted using various agricultural plant residues as organic additives. In the selection of organic materials, it is essential to be sustainable as well as looking at the waste of a specific process.

In this study, pine nutshell, which was produced as waste during pine nut production, was used as an organic additive. In the production stage, the amount of peanut and shell occur in approximately the same proportions. Furthermore, the easy grinding of the shell due to its brittle structure increases its potential for use as a pore former. The thermal conductivity value of the fired brick depends on the density of the brick, the shape and amount of the pores. These properties are directly related to the mineralogical properties of the clay used, as well as the firing properties of the brick. Temperature and time are the most critical parameters in the firing process of the brick. In porous brick production, besides these two parameters, size distribution and proportion of additive are other important parameters.

In this study, thermal conductivity properties of fired bricks, the weight ratio of additive, firing temperature, and firing time experimental parameters were determined by Box-Behnken experimental design and their relationships were investigated.

Definition of the Problem

The heat transfer behavior of microporous materials can be described as, employing convection, radiation, and conduction mechanisms [14]. Thermal conductivity is a valuable material property that plays a crucial role in the heat transfer calculations of the construction materials with the aim of heat insulation [15]. It is defined as the amount of heat flow under the unit temperature gradient for a unit area that includes some or all of the heat transfer mechanisms [14]. Thermal conductivity of the brick material depends on the structure, density or porosity, and the temperature. It is especially sensitive to porosity which scatters the thermal flow [3]. The presence of porosities in the brick structure leads to a decrease in thermal conductivity. In this study, C-Therm TCi Thermal Conductivity Analyzer based on Modified Transient Plane Source (MTPS) technique was used to characterize the thermal conductivity of brick samples with micro-porosity. This device utilizes a single-sided, interfacial heat reflectance sensor that applies a momentary constant heat source to the sample. The measurement is high-speed. Thermal conductivity value is measured directly, and a detailed overview of the heat transfer properties of brick samples is provided. The American Society for Testing and Materials (ASTM) standard D7984 defines the complete method [13].

In the present study, experimental data which uses pine nutshell as a pore making additive in the brick is considered. Firing temperature, firing time, and amount of additives were determined as experimental design parameters. The 17 experimental sets were determined based on the Box-Behnken experimental design, and the thermal conductivity values of the fired bricks were measured with C-Therm TCi Thermal Conductivity Analyzer. In the optimization part of the study, objectives, constraints, and design variables of the optimization problems are carefully defined and are listed in Table 16.1. Problem 1 design parameters were determined so that there was no restriction between the maximum and minimum test parameters used in production. Problem 2 is determined by the firing parameter values that can be used in experimental studies. In line with these constraints, X_i and X_k are variables with an accuracy of one-thousandth, and X_j is an integer.

Table 16.1 Objective functions, constraints, and design variables of the optimization problems.

Problems	Objectives	Constraints	Design Variables
1	Thermal conductivity coefficient λ (W/mK) minimization	Pine nuts addition (wt%), $(0 < X_i < 10)$ Firing temperature (°C), $(850 < X_j < 1{,}050)$ Firing time (hour), $(2 < X_k < 6)$	Pine nuts addition (X_i) Firing temperature (X_j) Firing time (X_k)
2	Thermal conductivity coefficient λ (W/mK) minimization	Pine nuts addition (wt%), $X_i \in \{0, 5, 10\}$ Firing temperature (°C), $X_j \in \{850, 950, 1{,}050\}$ Firing time (hour), $X_k \in \{2, 4, 6\}$	Pine nuts addition (X_i) Firing temperature (X_j) Firing time (X_k)

Regression Analysis

In this study, pine nutshell was used as a pore making inorganic additive in porous brick production. In the production of bricks, the number of additives, firing temperature, and duration of the bricks was determined as experimental parameters. The sets to be applied in the experimental studies were determined by Box-Behnken experimental design sets in three levels of three production parameters.

The parameters in the experimental sets and the thermal conductivity values of the produced bricks are given in Table 16.2. The second-order nonlinear polynomial objective function was introduced in order to define the effects of experimental parameters on the thermal conductivity of bricks. In the preparation of objective function, regression analysis and ANN were used together in the form of "neuro-regression". For this purpose, 13 of the experimental sets were used as training data in

Table 16.2 Experimental parameters and average thermal conductivity values of fired clay bricks.

Data	Run Numbers	Pine Nut Shell Addition (wt%)	Firing Temperature (°C)	Firing Time (h)	Thermal Conductivity (W/mK)
Training	1	0	1,050	4	0.8150
	2	0	850	4	0.6413
	5	5	850	2	0.5249
	6	5	1,050	2	0.6110
	7	5	850	6	0.5323
	8	5	1,050	6	0.6326
	9	5	950	4	0.5265
	10	5	950	4	0.5237
	12	5	950	4	0.5106
	13	5	950	4	0.5340
	14	10	1,050	4	0.5603
	16	10	950	6	0.5470
	17	10	950	2	0.4938
Testing	3	0	950	6	0.8093
	4	0	950	2	0.6854
	11	5	950	4	0.5637
	15	10	850	4	0.4989

the formation of the objective function, and R^2 value was calculated by regression analysis (R^2 training). The remaining four sets of experiments were used to test the prediction ability of the objective function, prepared as testing data, and then the predictive ability was determined statistically (R^2 testing).

The general form of regression models introduced for training data is as follows:

$$Y = a_0 + a_1 X_i + a_4 X_i^2 + a_2 X_j + a_7 X_i X_j + a_5 X_j^2 + a_3 X_k + a_8 X_i X_k + a_9 X_j X_k + a_6 X_k^2 \tag{16.1}$$

where X_i is the parameter for the weight percent of pine nutshell addition, X_j represents firing temperature, and the parameter firing time is denoted by X_k.

The objective function of the optimization problem, the polynomial model, given in Equation 16.2, was obtained using the training data (see Table 16.2). The $R^2_{training}$ value of the model was calculated as 0.92. By using the test data, $R^2_{testing}$ value was found to be 0.73 when the differences between the real (experimental) values and the predicted values were taken into consideration. A high R^2 value during the testing step indicates that the prescribed model is appropriate and describes the engineering problem well. Furthermore, the fact that the prepared objective function defines a three-variable engineering problem with only ten terms increases both its intelligibility and usability.

$$Y = 3.593 + 0.0224\ X_i + 0.0029X_i^2 - 0.0068\ X_j - 0.00008\ X_i\ X_j + 0.000004X_j^2$$
$$- 0.036\ X_k + 0.0019\ X_i\ X_k + 0.000017\ X_j\ X_k + 0.0016\ X_k^2 \qquad (16.2)$$

Optimization of Thermal Conductivity of Micro-porous Clay Bricks

Minimization of thermal conductivity in the production of porous bricks was realized with four different optimization algorithms for two problems, including discrete and continuous domain constraints. The results obtained from the solution of both problems are shown in Table 16.3.

It is observed that all the thermal conductivity values were obtained as 0.4874 W/mK, except the result for Problem 2 based on the RS algorithm (0.4888 W/mK). It is foreseen that the higher firing time will increase the sintering amount of the bricks and increase the bulk density. Since bulk density and thermal conductivity are inversely related, the lowest firing time of 2 hours was found in all solutions. Although the firing temperature is considered as a parameter which can be evaluated as the firing time, it has been determined in experimental studies that a temperature of 950°C is required to burn the additive by leaving a space in a place altogether. In

Table 16.3 Optimum predicted thermal conductivity values and experimental parameters of porous clay bricks.

Optimization Algorithms	Problems	Thermal Conductivity Values (W/mK)	Pine Nut Shell Addition (wt%)	Firing Temperature (°C)	Firing Time (h)
Differential Evolution	1	0.4874	8.408	930.495	2
	2	0.4874	8.400	930	2
Nelder-Mead	1	0.4874	8.408	930.497	2
	2	0.4874	8.420	931	2
Random Search	1	0.4874	8.408	930.495	2
	2	0.4888	8.410	949	2
Simulated Annealing	1	0.4874	8.409	930.513	2
	2	0.4874	8.34	932	2

the optimization studies, values between 930°C and 949°C were found to support this preliminary information.

Conclusion

Fired brick is known as one of the most crucial wall elements of building materials, and it must have low thermal conductivity values. For this purpose, the materials have been produced in the porous form. It is critical to determine production parameters that provide optimum values. Stochastic optimization algorithms are used to solve the design problems on the production of fired clay-based brick construction materials. The effects of the amount of pore making additive materials and the firing regime on the thermal conductivity of fired porous bricks were also investigated.

The experimental results of fired brick production are examined; the lowest thermal conductivity value was seen as 0.4938 W/mK. This value is obtained also in the 17th experiment, where the additive amount is 10%, the firing temperature is 950°C, and the firing time is 2 hours. The second-lowest value was obtained in the 15th experiment, where the additive amount was 10%, the firing temperature was 850°C, and the firing time was 4 hours. According to these two results, it can be said that firing time is more effective on thermal conductivity than firing temperature. Furthermore, the high additive amount, the low firing temperature and the time can be said to be the reason for the low density of the brick and therefore, the low thermal conductivity.

In the modeling part, the best fit with the training results was obtained with the nonlinear second-order polynomial model with high R^2 value as 0.92. This model is a highly usable and appropriate model due to its low component and proper description of this engineering phenomenon. In the optimization section, it is interesting that all optimization algorithms gave similar results in both problems. Minimum thermal conductivity value for Problems 1 and 2 was found to be 0.4874 W/mK. In all the results, the firing time was 2 hours as in the experimental results. However, although experimental data led us to an additive amount of 10% and the firing temperature to 850°C, the amount of pore-making additive was found to be optimal at 8.4% and 930°C.

References

[1] Zhang, L. 2013. Production of bricks from waste materials—A review. Constr. Build. Mater. 47: 643–655.
[2] Velasco, P.M., Ortíz, M.M., Giró, M.M. and Velasco, L.M. 2014. Fired clay bricks manufactured by adding wastes as a sustainable construction material—A review. Constr. Build Mater. 63: 97–107.

[3] Kornmann, M. 2007. Clay bricks and roof tiles, manufacturing and properties. Société de l'industrie Minérale, Paris.

[4] Chen, Y., Zhang, Y., Chen, T., Zhao, Y. and Bao, S. 2011. Preparation of eco-friendly construction bricks from hematite tailings. Constr. Build. Mater. 25(4): 2107–2111.

[5] Lingling, X., Wei, G., Tao, W. and Nanru, Y. 2005. Study on fired bricks with replacing clay by fly ash in high volume ratio. Constr. Build. Mater. 19(3): 243–247.

[6] Dondi, M., Marsigli, M. and Fabbri, B. 1997. Recycling of industrial and urban wastes in brick production—A review. Tile Brick Int. 13(3): 218–225.

[7] Segadaes, A.M. 2006. Use of phase diagrams to guide ceramic production from wastes. Adv. Appl. Ceram. 105(1): 46–54.

[8] Sutcu, M. and Akkurt, S. 2009. The use of recycled paper processing residues in making porous brick with reduced thermal conductivity. Ceram. Int. 35(7): 2625–2631.

[9] Sutcu, M., del Coz Díaz, J.J., Rabanal, F.P.Á., Gencel, O. and Akkurt, S. 2014. Thermal performance optimization of hollow clay bricks made up of paper waste. Energy Build. 75: 96–108.

[10] Sutcu, M., Alptekin, H., Erdogmus, E., Er, Y. and Gencel, O. 2015. Characteristics of fired clay bricks with waste marble powder addition as building materials. Constr. Build. Mater. 82: 1–8.

[11] Sutcu, M., Ozturk, S., Yalamac, E. and Gencel, O. 2016. Effect of olive mill waste addition on the properties of porous fired clay bricks using Taguchi method. J. Environ. Manage. 181: 185–192.

[12] Menezes, R.R., Neto, H.M., Santana, L.N.L., Lira, H.L., Ferreira, H.S. and Neves, G.A. 2008. Optimization of wastes content in ceramic tiles using statistical design of mixture experiments. J. Euro. Ceram. Soc. 28(16): 3027–3039.

[13] ASTM International, ASTM Standard D7984, Standard Test Method for Measurement of Thermal Effusivity of Fabrics Using a Modified Transient Plane Source (MTPS) Instrument, ASTM International, West Conshohocken, PA, 2016.

[14] Bejan, A. and Kraus, A.D. 2003. Heat Transfer Handbook. John Wiley and Sons, New York.

[15] Bhattacharjee, B. and Krishnamoorthy, S. 2004. Permeable porosity and thermal conductivity of construction materials. J. Mater. Civ. Eng. 16(4): 322–330.

Index

T - #0423 - 071024 - C236 - 234/156/11 - PB - 9780367510022 - Gloss Lamination